Independent Science Challenges

Fascinating science projects to challenge and extend able students

scientific hoaxes

global warming

nuclear power

cloning

Charlotte Samiec

HAWKER BROWNLOW EDUCATION

Cover & Book Designer: Lauren Mills

Published in Australia by

HAWKER BROWNLOW
EDUCATION

P.O. Box 8580, Heatherton,
Victoria 3202, Australia
Phone: (03) 8558 2444 Fax: (03) 8558 2400
Toll Free Ph: 1800 33 4603 Fax: 1800 15 0445
Website: http://www.hbe.com.au
Email: orders@hbe.com.au

Printed in Australia

Code: HB6177
ISBN: 1 74101 617 7
0306

Contents

Introduction

What are *Independent Science Challenges*?

Independent Science Challenges are long-term, in-depth, open-ended science research projects designed to challenge and extend gifted upper-primary students.

Why are *Independent Science Challenges* great learning experiences?

Independent Science Challenges are great learning experiences because they:

- are **student-centred** – catering for different abilities, learning styles and personal interests
- are **inquiry-based** – requiring a discriminating search for information using a variety of media as well as consultation, experimentation and observation
- are **open-ended** – defying simplistic answers and placing no limitations on where the student's interest, research and effort can take them
- are **independent** – allowing the student to take responsibility for their own learning unrestrained by the collective pace of a group or detailed direction from a teacher
- involve **higher-order thinking** – using analysis, synthesis and evaluation of information concerning complex and controversial topics
- provide opportunities for **creativity** in original and imaginative products and presentations
- are **long-term** – necessitating perseverance and effort over many weeks or months
- are about **people** – discovering positive role-models and possible career-paths as students learn about and meet scientists
- help develop **responsible citizenship** – providing the student with practice in problem-solving, debating, leadership, organisational skills and informed and rational decision making; encouraging the student to consider the use and misuse of science; to acknowledge the need for ecological sustainability; to take personal responsibility and act for the welfare of Earth and all living things
- are **inter- and intra-disciplinary** – drawing on many disciplines of science and incorporating other subjects such as maths, English, history, geography, art and music
- are **supportive of learning** – providing a scaffold within which the student has freedom to pursue interests, make choices about what to research and how to demonstrate learning
- require **advanced and critical reading skills.**

Why are *Independent Science Challenges* an optimal way to learn Science?

Independent Science Challenges are an optimal way to learn Science because they:

- encourage curiosity and creativity
- are concerned with real-life scientific issues and controversies necessitating dialogue with scientists and practical learning outcomes demonstrated for real-world audiences
- are hands-on – involving experimentation, construction, observation, use of data and meaningful use of Information Communication Technologies
- involve asking good questions, planning and conducting investigations and effectively communicating results
- require understanding and use of scientific methods, scientific language and safe working practices
- develop attributes of scientists – scientific literacy, the ability to access and evaluate information, effective use of technologies, respect for truth, healthy scepticism and objectivity
- develop positive attitudes towards science – appreciating scientific achievements, the important role of science and the contribution of scientists from many cultures

Are *Independent Science Challenges* suitable for all students?

Just as students with lower academic ability need special provision and adjustment to the curriculum, students with high or extraordinary ability also require the curriculum to be differentiated in terms of:

- advanced, complex and challenging content
- allowances for self-directed, in-depth learning, higher-order and divergent thinking and creative problem solving
- high-level, real-life or life-like products
- learning environments that provide for independence, allow flexibility and choice.

While most students benefit from the opportunity to participate in an independent study, *Independent Science Challenges* are intended for very able upper primary/lower secondary school students and may need modification or significant teacher support to be used with average-ability students.

How do I know which students would benefit from participating in an *Independent Science Challenge*?

There are many useful check-lists available to help you identify gifted students. In brief, some of the characteristics you may see in students are:

- excellent general and/or specific knowledge – 'the walking encyclopaedia'
- intense curiosity
- prolonged attention span
- tendency to become immersed in or obsessed with particular topics
- advanced reading ability
- preference to work by themselves – can be unenthusiastic about 'group work' or bossy with others
- preference for the company of the teacher, adults or older children – may not socialise well with chronological peers
- responsibility – displays maturity
- ability to do well in tests with minimal effort – may not pay attention but still know all the answers
- intense emotions – empathises with others and is concerned with the injustices of the playground or world
- advanced sense of humour – enjoys jokes and puns
- creativity, originality and imagination
- problem-solving ability – enjoys puzzles and strategy games
- advanced vocabulary
- dislike for unchallenging, routine work – works best when challenged
- may be disruptive when bored – the 'class clown' or teacher-baiter.

(Note: Some gifted girls disguise their abilities to blend in with their age-peers.)

What is the teacher's role if the science challenge is 'independent'?

The teacher (or mentor):

- recognises the student's strengths, weaknesses and learning style(s) and helps them choose appropriate learning experiences that challenge and extend
- takes an interest in the student's learning – discussing, encouraging and guiding
- establishes limits, standards and safety guidelines
- is flexible and allows the student the space, time and facilities to do their work
- assists the student to find and access resources (including human resources)
- provides feedback and constructive criticism
- doesn't pretend to have all the answers, but instead, provides a role-model of the 'life-long learner'.

How to use *Independent Science Challenges*

Independent Science Challenges can be used:

- for enrichment and extension with individual gifted students in a mixed-ability class
- as a challenging activity for gifted students in a 'pull out' program
- as part of the curriculum for all students in a gifted class
- for keen science students who want ideas for science fairs and competitions
- as a resource for 'home schooling'
- (with modification) for small groups or mixed-ability classes.

Step 1

Independent Science Challenge topics can be decided upon through negotiation between students and their teacher/mentor and parents. Considerations are likely to include curriculum priorities, personal interests, availability of local resources, specialists and sites of scientific interest.

Students will need a copy of:

- their particular *Independent Science Challenge*
- *Pre-launch check – Choices* (pp. 13–14)
- *Portfolio possibilities – Choices* (pp. 16–20)
- *The big finish – Ideas* (pp. 22–23)
- *Science challenge assessment rubric* (at teacher's discretion – p. 25)
- *Tips for students* (p. 26)

Step 2

Students begin by completing a *Pre-launch check* (see pp. 12–14). This is an opportunity for the student to consider and communicate what ideas they have and what information they already know prior to commencing research.

Step 3

Students commence research by attempting to answer the *Starter questions.* These questions are designed to give the student an introduction to the basic facts, concepts, vocabulary and interesting and controversial aspects of the topic.

Step 4

Students pursue aspects of the topic that particularly interest them by completing a number of activities (see *Portfolio possibilities* pp. 15–20). The student will need to negotiate with their teacher or mentor regarding how, when and the number of activities that will need to be completed for assessment.

Step 5

The *big finish* (see pp. 21–23) is a culminating activity where the breadth and depth of the student's learning can be showcased.

The format of the challenges

Title

Introduction

Global Warning?

The world is getting hotter. Plants and animals are becoming extinct as their environments change. Floods, heat waves, hurricanes, storms and forest fires are increasing in frequency and severity. Glaciers and icebergs are melting, the ocean is expanding and already some countries are beginning to disappear under the waves.

What have we done to cause global warming and what will the consequences be?

The challenge in brief

Pre-launch check:

What will happen if temperatures on Earth continue to rise? Create a Consequences Chart.

Alternative ideas? See pages 13–14

What does the student know before commencing research

Space to write down other pre-launch ideas

Starter questions:

How does nature's 'greenhouse effect' work and how has human activity enhanced it?

How does nature's 'carbon cycle' work and how has human activity interfered with it?

What evidence is there that climate ch

How can such a small change in tempe

Questions designed to help the student understand the basics of the topic

How might global warming affect life on Earth in the future?

How do scientists gather information/evidence about Earth's past and present climate and how are they able to predict what is likely to happen to our planet?

What should we be doing now to prevent global warming becoming a catastrophe?

The format of the challenges

A selection of activities to choose from to learn more about the topic

Portfolio possibilities:

- Find and/or design and conduct experiments to show how the greenhouse effect and the consequences of global warming work. (Include experiments dealing with the thermal expansion of water.)
- Investigate the properties of ice and conduct experiments to discover what the melting of ice at the poles will do to rising sea levels.
- After researching the carbon cycle, write a script and perform 'This is your 4-billion year life!' about what might happen to a carbon atom during its existence on Earth.
- Use the Internet to find weather data for your region for the last 100 years. Analyse the data to see if you find evidence of significant climate change.
- Report on an endangered animal that is further threatened by climate change.
- Write the front page of 'The Polar ExDepress' – a newspaper about the effects of climate change as experienced at Earth's Arctic and Antarctic regions.
- Note evidence of climate change on a map of the world.

Alternative ideas? See pages 16–20

Space to write down other portfolio possibilities

Culminating activity

The big finish:

Educate class members about the issues of global warming and organise for them to calculate their families' annual carbon dioxide emissions. In ratifying the Kyoto Protocol, countries around the world have agreed to reduce their greenhouse emissions. In the same spirit, encourage class members to examine how they can reduce or counteract their own contributions to carbon dioxide emissions and sign up to a mini-Kyoto Protocol.

Alternative ideas? See pages 22–23

Space to write down other big finish ideas

Pre-launch check – An introduction

The *Pre-launch check* section of the science challenge is an opportunity for the student to consider and communicate what ideas they have and what information they already know prior to commencing research.

This kind of activity is valuable because:

- it stimulates questions and interest and helps engage the student in the topic
- it can be used as a 'baseline' to measure subsequent learning against and is therefore useful for assessment purposes
- it helps the teacher/mentor be aware of what prior knowledge the student brings to the topic. In the case of very able students this can be quite significant and even, for the 'teacher', a little surprising.

If a student demonstrates that they already have a very good understanding of the challenge – topic, expectations, time-limits, activities etc. – the challenges can be differentiated appropriately to avoid the situation where they are wasting time 'learning' what they have already mastered.

A suggestion for a possible *Pre-launch check* is included in each science challenge. Alternative ideas are listed on the following pages and should be made available to the student.

Pre-launch check – choices

A pre-launch check suggestion has been included in your science challenge. Alternatively, select from the list below or use your own ideas.

Note: All these choices relate specifically to the topic of your science challenge – not to science generally.

Generate a mind map (concept-map/web).

Try a test paper from a higher year level.

Make a list of facts:

- on a long piece of paper like a roll from a cash register – how many metres can you write?
- on links of a paper-chain – how long can you make the chain?
- on small paper squares and glue them down – how tall a column can you make?

List:

- 100 words associated with …
- 10 controversial aspects of …
- 10 questions that can be answered using the main word of the topic, for example if you are researching magnetism, what are 10 questions that can be answered with 'magnetism'?
- 10 ways your topic has changed or affects the world.

Fill a page or whiteboard with drawings, diagrams or facts.

Do a PMI (Plus, Minus, Interesting) Analysis.

Do a SWOT (Strengths, Weaknesses, Opportunities, Threats) Analysis.

Complete an A–Z of facts. Write the alphabet down one side of a page and write one fact you know for each letter.

Make an illustrated timeline.

Draw a labelled diagram.

Use the main topic word and make an acrostic by writing one fact or concept for every letter.

Try answering the *Starter questions* before you do any research.

What if — vanished/stopped working/didn't exist? What would the consequences be?

Imagine:

- you are abducted by aliens who know nothing about your topic – what would you tell them?
- you own a shop that sells everything to do with your topic – list what you sell
- your topic is a person and they have agreed to talk at your school assembly – write their speech.

Use the Internet and log on to a reputable science discussion forum. Start and participate in a discussion using the information you already know.

Grab a glossary from a book or website about your topic. Copy the words from the glossary and try to explain their meanings.

Make a before/after or with/without picture.

See how long you can talk about your topic before you run out of knowledge.

Collect questions from classmates and family. Try to answer them.

Find and try some online quizzes about your topic and note your strengths and weaknesses.

Portfolio possibilities – An introduction

By researching and answering the *Starter questions* the student will have learnt a great deal of general information about their topic. They are now ready to choose several activities from the *Portfolio possibilities* section of the science challenge or from the alternative ideas listed on the following pages.

Completing these activities will enable the student to:

- deepen and broaden their understanding of the topic using their preferred learning styles
- specialise in and pursue aspects of the topic that they find particularly interesting
- use their individual talents to communicate their learning to others
- try a variety of new and exciting ways of learning and communicating
- access scientists and other professionals; participate in real-world scientific activities and visit places where scientists work
- think critically about the use and misuse of science and technologies, points of view, bias and the quality and reliability of information sources
- be creative and produce new ideas and products.

The student will need to negotiate with their teacher or mentor regarding how and when these portfolio activities will need to be completed for assessment.

Specific *Portfolio possibilities* are included in each science challenge. Alternative ideas are listed on the following pages and should be made available to the student.

Portfolio possibilities - choices

Several *Portfolio possibilities* have been included in your science challenge. Alternatively, select from the list below or use your own ideas.

Note: All these choices relate specifically to the topic of your science challenge – not to science generally.

Find and/or design and perform experiments related to your topic.

Recreate famous or historical experiments.

Demonstrate an experiment to an audience explaining the proceedings and outcome.

Use the Internet or computer software to participate in 'virtual' experiments.

Collect:

- examples of ...
- places where ...
- reasons why ...
- dangers associated with ...
- recipes for ...
- experiments to show ...
- arguments for or against ...
- information about famous scientists associated with ...
- consequences of ...
- amazing facts about ...
- ways to ...
- newspaper and magazine articles about ...
- statistics showing ...
- questions about ...
- lists of scientific careers associated with ...
- eye-witness accounts of ...
- evidence for and/or against ...
- pictures of ...

words associated with …

solutions to …

uses for …

Refute an unsound argument or unscientific opinion with facts.

Analyse your topic using Edward de Bono's Six Thinking Hats.

Design an invention that …

Construct or deconstruct something.

Write:

a report suitable for publication in a science magazine or school newsletter

an imaginative newspaper story – historical, contemporary or futuristic

a review of a book, website, television program or film

a story about a scientist, discovery or invention

a quiz

a biography

a poem

an essay

an imaginative or realistic journal of someone involved in or affected by …

a list of other *Portfolio possibilities* you would like to try if you had more time

10 questions about your topic that scientists still can't answer

a letter to a scientist, newspaper or politician

a fictional interview based on real facts

an obituary for a significant scientist or even a defunct idea

a prediction based on current facts

an observation

a personal reflection.

Compare and contrast:

- now and then
- opinions
- discoveries
- scientists
- theories
- technologies
- popular fiction with reality
- old and new ideas
- methods
- myths and facts.

Create:

- a 'Hall of Fame' of scientists who have achieved great things
- an annotated list of informative websites (with links), articles or books
- a timeline showing the development of scientific understanding about …
- an interactive learning centre
- a catalogue
- a flow chart that explains …
- a glossary
- a show based on a popular game show format
- a concept (mind) map.

Work out the maths to ….

Perform:

- a rap
- a dance
- a role play, play, skit, monologue or puppet show.

Take photographs.

Generate new ideas using the SCAMPER (Substitute, Combine, Adapt, Modify/Magnify, Put to other uses, Eliminate, Reverse) thinking strategy.

Keep:

- a journal (learning log)
- an annotated scrap book of relevant news items.

Make:

- a slide-show or PowerPoint presentation
- a television or radio advertisement or program
- a video
- a map
- a diorama
- a prediction
- a working model
- a mural
- a mobile
- a poster
- a collage
- a comic strip
- a documentary suitable for radio.

Examine something under a microscope or with a telescope.

Draw or write 'before and after' descriptions or pictures.

Give an informative and entertaining talk to your class, an older or younger class or to a community group.

Design and teach others to play a game based on your topic.

Do a SWOT (Strengths, Weaknesses, Opportunities, Threats) Analysis.

Prepare and conduct a survey or questionnaire. Analyse the results.

Compose a soundtrack.

Visit a relevant:

- museum exhibition
- factory
- laboratory
- clinic
- observatory
- geological site
- man-made environment
- natural environment.

Propose new ideas, new uses for and/or new ways of doing things.

Grow/breed, observe and measure something living.

Teach someone something about your topic.

Investigate a claim that seems unscientific and find out the truth.

Research, prepare questions and interview:

- a scientist or other professional
- someone who has experienced ...
- an avid supporter or opponent of ...

Prepare a pamphlet containing useful information about your topic.

Take part in an actual scientific survey to monitor, observe or record something.

Draw a diagram.

Paint a picture.

Explore old books, movies, libraries, for example, looking for outdated scientific ideas.

Use the Internet to:

- find real-life/real-time data
- ask questions of scientists
- take part in an online discussion/forum.

The big finish – An introduction

The big finish section of the science challenge is a culminating activity that provides an opportunity for the student to demonstrate their grasp of the complexities of their topic and communicate their learning to a real-world audience.

The big finish is:

- a culmination and celebration of months of hard work
- not just a summary of regurgitated facts but involves translating a deep knowledge of, and original ideas about, a complex topic into a product with a specific purpose
- an opportunity to communicate scientifically about a 'real-world' issue to a real audience
- an opportunity for the student to mirror the work of adults/professionals
- an opportunity to develop citizenship, leadership and organisational skills
- an opportunity to make a real difference towards a just and sustainable world
- something for the student and teacher/mentor to discuss well ahead of time so that necessary preparations can be made and an appropriate audience found.

A specific *big finish* activity is included in each science challenge. Alternative ideas are listed on the following pages and should be made available to the student.

The big finish – Ideas

A *big finish* suggestion has been included in your science challenge. Alternately, select from the list below or use your own ideas.

Note: All these choices relate specifically to the topic of your science challenge – not to science generally.

Become very familiar with a museum's collection and displays. Create an interesting and challenging worksheet for visitors to the museum to complete.

Research, design and conduct detailed, (and possibly) long-term experiments. Enter your findings and conclusions in a science competition.

Organise an information evening for your local community. Invite speakers with a variety of viewpoints and give an address yourself on an aspect of the topic that particularly interests you.

Critically evaluate one or more websites or books that present incorrect or misleading information. Citing reliable scientific sources, highlight the errors. Write or email your opinions to the author(s) of the websites or books.

Organise your class or school to participate in a relevant national or global campaign or action day. Provide publicity, information and activities.

Make:

- a documentary-style television or radio program
- an interactive in-depth website
- a PowerPoint presentation for a specific audience.

Compose the soundtrack for an informative pictorial presentation. Because it uses images and no words, you must convey meaning of the presentation with your choice of music.

Conduct interviews/questionnaires to ascertain peoples' knowledge, questions and fears about a controversial aspect of your topic. Research, write and distribute a question and answer-style pamphlet citing up-to-date scientific responses to their questions and worries.

Contact the producers of a local children's television program or children's section of a newspaper and convince them to broadcast/publish an interesting and important aspect of your research.

Teach your topic to your class, a younger group, parents or teachers using a variety of methods.

Write:

a letter to someone who matters (and who has the power to change things) detailing your opinion and ideas

an essay and find a qualified person to read, discuss and assess it

a report for publication in a science magazine or school or community newsletter

a well-researched and scientifically-sound fiction story

a play and then produce and perform it

the script for a children's television science show.

Participate in:

a debate about a controversial aspect of your topic

a 'mock trial'

work-experience or work-shadowing with a scientist

a club and lead an activity.

Create:

a large and informative mural for a particular place and audience

a complex and informative game that will entertain and educate

a WebQuest

a detailed model

a hands-on display or learning centre packed with information designed to appeal to a variety of learners

an information book or comic

a new product

a display for your school/public library or shopping centre

an entertaining and informative quiz and variety show for your class.

Put on a dramatic performance in which you role-play a diverse range of characters or actual people with a variety of points of view and personal perspectives. Use simple but significant props.

Assessing *Independent Science Challenges*

An *Independent Science Challenge* is a lengthy and complex undertaking and the teacher/ mentor needs to be involved in ongoing assessment to ensure that the student:

- has helpful guidance and encouragement
- is organised and makes good use of time
- is on the right track – knowing how and where to find suitable information
- is completing work of an appropriate standard – neither coasting nor getting out of their depth
- has information collected about them that can be used by parents and teachers to make good decisions about future activities and educational provision
- is not (as is the case of many 'perfectionist' gifted children) wearing themselves out

Some gifted students may produce portfolio tasks and a *big finish* of such a high academic standard that it may be beyond the ability of their chronological peers and even their teacher to judge. It would be preferable for these students to be able to work with a knowledgeable mentor and to present to an audience of scientists and/or professionals (or at least well-prepared adults) so that the extent of their work can be fully appreciated.

Some useful tools for assessment include:

- a rubric of pre-established criteria (see p. 25)
- a learning journal (a personal record of questions, thoughts and progress)
- self-evaluation proforma
- question and answer session after a presentation
- interviews
- feedback (using an evaluation proforma) from audiences of peers, parents and/or professionals
- awards, marks and comments from judges (if the student has entered their work in a science competition)
- the P*re-launch check* which can be redone at the completion of the science challenge.

Science challenge assessment rubric

Criteria		A		B		C
Responces to starter questions	**A**	Used a wide variety of up-to-date sources to complete all *Starter questions*. Answers were detailed, clearly expressed in the student's own words and demonstrated an excellent understanding of all the relevant facts and concepts.	**B**	Used a variety of up-to-date sources to complete all *Starter questions*. Answers were clearly expressed in the student's own words and demonstrated a good understanding of all the relevant facts and concepts.	**C**	Used a number of up-to-date sources to complete all *Starter questions*. Answers were usually clearly expressed in the student's own words and demonstrated a satisfactory understanding of all the relevant facts and concepts.
Effort	**A**	Completed 8 or more portfolio tasks and a *big finish* culminating activity.	**B**	Completed 6 or more portfolio tasks and a *big finish* culminating activity	**C**	Completed 4 or more portfolio tasks and a *big finish* culminating activity.
Use of resources	**A**	Used a wide variety of electronic, written and human resources including interviews with at least 2 scientists or professionals and excursions to at least 2 relevant locations.	**B**	Used a wide variety of electronic, written and human resources including an interview with at least 1 scientist or professional and an excursion to at least 1 relevant location.	**C**	Used a wide variety of electronic, written and human resources including an interview with a scientist or professional or an excursion to a relevant location.
Comprehension and understanding	**A**	Consistently demonstrated a superior understanding of the scientific facts, concepts, issues and vocabulary involved.	**B**	Consistently demonstrated a good understanding of the scientific facts, concepts, issues and vocabulary involved.	**C**	Demonstrated a satisfactory understanding of the scientific facts, concepts, issues and vocabulary involved.
Use of high level thinking skills	**A**	Always demonstrated high-level thinking skills (application, analysis, evaluation, synthesis) in the completion of portfolio tasks.	**B**	Usually demonstrated high-level thinking skills (application, analysis, evaluation, synthesis) in the completion of portfolio tasks.	**C**	Sometimes demonstrated high-level thinking skills (application, analysis, evaluation, synthesis) in the completion of portfolio tasks.
Use of critical thinking	**A**	Consistently demonstrated an ability to critically evaluate information for relevance, accuracy and bias.	**B**	Usually demonstrated an ability to critically evaluate information for relevance, accuracy and bias.	**C**	Sometimes demonstrated an ability to critically evaluate information for relevance, accuracy and bias.
Working scientificaly	**A**	Completed a large number of complex and/or lengthy experiments and/or observations in a scientific manner.	**B**	Completed several complex and/or lengthy experiments and/or observations in a scientific manner.	**C**	Completed several experiments and/or observations in a scientific manner.
Habits of mind	**A**	Consistently demonstrated the attributes of a successful scientist – curiosity, enthusiasm, scepticism, perseverance and methodicalness.	**B**	Usually demonstrated the attributes of a successful scientist – curiosity, enthusiasm, scepticism, perseverance and methodicalness.	**C**	Demonstrated some of the attributes of a successful scientist such as curiosity, enthusiasm, scepticism, perseverance and methodicalness.
Creativity	**A**	Demonstrated outstanding creativity in a wide selection of portfolio tasks	**B**	Demonstrated a high-level of creativity in a selection of portfolio tasks.	**C**	Demonstrated some creativity in a selection of portfolio tasks.
Responsibility and indpendence	**A**	Always demonstrated a responsible and independent attitude to use of time, resources and special provisions	**B**	Usually demonstrated a responsible and independent attitude to use of time, resources and special provisions.	**C**	Sometimes demonstrated a responsible and independent attitude to use of time, resources and special provisions.
Additional criteria (e.g. for controversial or enviromental topics)						
Tolerance, flexability and open-mindedness	**A**	Consistently demonstrated an ability to consider issues from a wide variety of points-of-view.	**B**	Usually demonstrated an ability to consider issues from a variety of points-of-view.	**C**	Sometimes demonstrated an ability to consider issues from a variety of points-of-view.
Citizenship	**A**	Demonstrated an exemplary commitment to environmental sustainability and responsible citizenship.	**B**	Demonstrated a commitment to environmental sustainability and responsible citizenship.	**C**	Demonstrated an interest in environmental sustainability and responsible citizenship.

Tips for students

Work safely at all times. Be very careful when conducting experiments that you are not exposing yourself or property to hazards. Get adult permission and assistance where appropriate. A science teacher may be able to help you with finding materials and equipment and choosing and conducting experiments.

The Internet is BIG. Your time is precious. Make sure you know:

- how search engines and the World Wide Web works
- how to identify key words (and synonyms) and use them to search
- how to use quotation marks (' '), and the plus (+) and minus (-) symbols to refine your search
- how to use the 'Find', 'Favourites', 'Back', 'Forward' and 'History' functions
- how to quickly cut, copy and paste using the Edit menu and shortcuts.

(If you are not sure, a teacher or mentor may be able to help or you could try learning using an online tutorial – enter 'Internet tutorial' into a search engine.)

Don't trust everything you find on the Internet. There is a lot of incorrect, worthless and biased material out there. When checking a webpage for information consider:

- Is it up-to-date? Can you see when it was last updated? Do the links work?
- Who wrote the information and/or set up the website? Is the author qualified? Is it a reputable organisation such as a university, museum or government agency or a company or individual wishing to push a biased view or sell you something?

When contacting scientists or other professionals remember that they are busy people. Usually they will be keen to help you if you are not wasting their time so:

- be an expert on the topic before you approach them
- don't ask questions you could easily find the answers to elsewhere
- research and ask questions specific to their field of expertise
- check workplace health and safety requirements regarding footwear etc.
- be punctual, neatly dressed and courteous
- if appropriate, ask if they would allow you to participate in or observe some laboratory or field work
- invite them along to your *big finish* event
- take your camera and politely request a photo so you can show your teacher/mentor/judge that you really did interview a scientist

Consider entering one of the many science competitions on offer. There are many to choose from that would be worthwhile big finishes for your *Independent Science Challenge*. Ask a science teacher or your state or national Science Teachers Association for details.

The Challenges

Global warning?

The world is getting hotter. Plants and animals are becoming extinct as their environments change. Floods, heat waves, hurricanes, storms and forest fires are increasing in frequency and severity. Glaciers and icebergs are melting, the ocean is expanding and already some countries are beginning to disappear under the waves.

What have we done to cause global warming and what will the consequences be?

Pre-launch check:

What will happen if temperatures on Earth continue to rise? Create a Consequences Chart.

Alternative ideas? See pages 13–14

Starter questions:

How does nature's 'greenhouse effect' work and how has human activity enhanced it?

How does nature's 'carbon cycle' work and how has human activity interfered with it?

What evidence is there that climate change is really occurring?

How can such a small change in temperature have such drastic effects?

How might global warming affect life on Earth in the future?

How do scientists gather information/evidence about Earth's past and present climate and how are they able to predict what is likely to happen to our planet?

What should we be doing now to prevent global warming becoming a catastrophe?

Portfolio possibilities:

- Find and/or design and conduct experiments to show how the greenhouse effect and the consequences of global warming work. (Include experiments dealing with the thermal expansion of water.)
- Investigate the properties of ice and conduct experiments to discover what the melting of ice at the poles will do to rising sea levels.
- After researching the carbon cycle, write a script and perform 'This is your 4-billion year life!' about what might happen to a carbon atom during its existence on Earth.
- Use the Internet to find weather data for your region for the last 100 years. Analyse the data to see if you find evidence of significant climate change.
- Report on an endangered animal that is further threatened by climate change.
- Write the front page of 'The Polar ExDepress' – a newspaper about the effects of climate change as experienced at Earth's Arctic and Antarctic regions.
- Note evidence of climate change on a map of the world.

Alternative ideas? See pages 16–20

The big finish:

Educate class members about the issues of global warming and organise for them to calculate their families' annual carbon dioxide emissions. In ratifying the Kyoto Protocol, countries around the world have agreed to reduce their greenhouse emissions. In the same spirit, encourage class members to examine how they can reduce or counteract their own contributions to carbon dioxide emissions and sign up to a mini-Kyoto Protocol.

Alternative ideas? See pages 22–23

'The cleanest, safest energy source there is.'

– Homer Simpson

Some say nuclear power plants are safe, clean and efficient producers of electrical energy capable of turning back the climate and environmental damage done by fossil fuels. Others say they are too risky – silently leaking harmful radiation, spawning waste that will be toxic for thousands of years. They say nuclear power plants are sullied partners in the dirty business of nuclear warfare and potential targets for terrorists. Nuclear power is highly controversial and it seems everyone has an opinion – but do you know the facts?

Is nuclear power too dangerous?

Pre-launch check:

Try answering the starter questions below to see how much you already know about nuclear power. How did you go? Now do some thorough research and see if you were right.

Alternative ideas? See pages 13–14

Starter questions:

How does a nuclear reactor work?

What's the difference between nuclear fission, nuclear fusion and a nuclear bomb?

What is radiation? Is it always dangerous?

How big and important is the world's current nuclear power industry?

What went wrong at the Three Mile Island (1979) and Chernobyl (1986) nuclear power plants?

What is nuclear waste? How is it managed and stored?

Portfolio possibilities:

- Make a model of a nuclear reactor. Explain to your class how it generates electricity.
- Prepare and interview (either in person or email) a scientist involved in the nuclear power industry about the work they do.
- Compare the effects on human health of air pollution from burning fossil fuels with the damage done by radioactive contamination from nuclear reactors.
- Prepare and conduct a survey to gauge public opinion and knowledge of nuclear power. Analyse your findings to determine how informed people really are.
- Complete a detailed SWOT Analysis of the world's nuclear power industry.
- Find ten links between nuclear power and nuclear weapons.
- Research current methods and ideas about the safe storage of radioactive waste from nuclear power plants. Come up with another scientifically sound idea.

Alternative ideas? See pages 16–20

The big finish:

Consider six characters with different experiences of, and opinions about, the benefits and dangers of nuclear energy (for example, a Greenpeace campaigner, a worker in a nuclear power plant, a Chernobyl victim, a member of the World Nuclear Association.) Using simple props, role-play each character and their point of view about nuclear power.

Alternative ideas? See pages 22–23

They came from the deep ...

For centuries sailors have told tales of monstrous waves rising suddenly from the deep and obliterating ships and their helpless crews. Now, science is revealing the awful truth about these legendary 'rogue' waves. On the 26th December 2004 over 300,000 people were washed away by a tsunami in the Indian Ocean.

What causes tsunamis and 'rogue' waves? How do scientists battle these monsters of the deep?

Pre-launch check:

Write every fact you know about waves on small squares of paper. Glue the squares down in the shape of a wave. How big a wave can you make?

Alternative ideas? See pages 13–14

Starter questions:

What is a wave?

How is a typical ocean wave formed? How does it move and change?

What causes 'rogue' waves? How are they different from typical ocean waves?

What causes tsunamis? How are they different from typical ocean waves?

How are 'rogue' waves and tsunamis similar and different?

How is information about 'rogue' waves and tsunamis collected and analysed?

What warning systems are in place and how do they work?

Portfolio possibilities:

- Make or borrow a wave tank and devise experiments to learn about waves. Record what you do and what you discover.
- Create a flow chart that explains the energy transfer in a tsunami or rogue wave.
- Collect accounts of 'rogue' waves and/or tsunamis and plot their locations on a map of the world. Determine which areas are most dangerous and why.
- Prepare questions and interview a maritime engineer and/or visit a maritime research facility.
- Using a model of an ocean liner or tanker explain what waves ships are designed to withstand and why they are often sunk when hit by a 'rogue' wave.
- Create a newspaper front page telling the true story of an encounter with a rogue wave or the impact of a tsunami.
- Present an illustrated talk on the likely consequences of a large asteroid plunging into the ocean.
- Research how waves can be used as a source of energy and make a model of a 'wave farm'.

Alternative ideas? See pages 16–20

The big finish:

Make an entertaining and educational radio program about waves.

Alternative ideas? See pages 22–23

Frankenfood?

For thousands of years humans have selectively bred crops and domestic animals to be bigger, hardier or tastier. Now, advances in biotechnology mean that scientists can directly alter the genes that make up our food and create new crops that can resist insect attack, grow in droughts or salty soil, and even repair damaged and poisoned environments. Scientists have developed fruit that can vaccinate us against, and cure, diseases; fast-growing fish and even chickens with four drumsticks may not be very far away. Genetically modified (GM) foods could save the lives or improve the health of millions of people BUT opponents claim that GM crops and animals are too dangerous – that they could ruin ecosystems and make us sick. Some say that tampering with genes is wrong and unnatural. Consumers all over the world are being urged to say no to 'Frankenfoods'.

Are genetically modified foods safe?

Pre-launch check:

How many genetically modified foods can you list that are already available in your supermarket? Can you explain how they were created and grown?

Alternative ideas? See pages 13–14

Starter questions:

How do cells work and reproduce?

What is DNA?

What do genes do?

What processes are used to genetically modify plants and animals?

What GM foods are already available?

What are the benefits of GM food-crops and animals?

What risks do GM food-crops and animals pose?

Portfolio possibilities:

- Make a model of a cell out of jelly and lollies. Use it to explain to your class how a cell works.
- Visit a biotechnology laboratory and make a photographic record of the GM process.
- Perform and document real and online experiments with cells. (Try extracting DNA for a start.)
- Trace the development of a common food-crop such as rice, wheat or corn from its original wild state to its modern agricultural and GM forms.
- Create diagrams showing how GM plants and animals could affect the biological diversity of particular environments.
- Investigate current precautions and testing regimes applied to GM foods – report what you have discovered using Edward de Bono's black and yellow thinking hats.
- Label a world map with details of 'pharming' projects being carried on around the globe. In each case consider who benefits from the research.
- Imagine it is 20 years in the future. Create a front page of a newspaper that commemorates the achievements and failings of the last 20 years of GM foods.

Alternative ideas? See pages 16–20

The big finish:

Organise a GM food information night in your local community. Invite speakers with a variety of viewpoints and give an address yourself on an aspect of the topic that particularly interests you.

Alternative ideas? See pages 22–23

A home with altitude

Way back in 1952, Werner von Braun, one of the world's first and foremost rocket engineers, predicted that 'development of the space station is as inevitable as the rising of the sun'. His dream has become a reality. Since the first two modules were launched in 1998 the International Space Station (ISS) has orbited the Earth sixteen times each Earth-day. Piece by piece this 400 kilometre-high laboratory has taken shape, becoming a 27,000 kilometre per hour home to astronauts from many different countries.

What is the role of the International Space Station and what is it like to live there?

Pre-launch check:

Can you fill a page with space station facts without repeating yourself?

Alternative ideas? See pages 13–14

Starter questions:

How has the ISS been put together?

How is the ISS different from its predecessors – Skylab and Mir?

What has the ISS achieved so far?

What is planned for the future of the ISS?

How are spacewalks to the outside of the ISS undertaken?

How do astronauts on the ISS work, eat, sleep, relax and go to the toilet?

Portfolio possibilities:

- Make a model of the ISS and use the model to explain to your class how the ISS works and what it does.
- Design a robot that would be useful on the ISS. Explain why.
- Imagine you were selected to visit the ISS. Write a journal based on the real experiences of space tourists.
- Use the Internet to find out if you can see the ISS from your location – watch it pass over!
- Analyse the arguments for and against spending money on exploring space by completing a PMI chart (Plus, Minus, Interesting/Implications).
- Find out who has been there and for how long. Compile a list of ten interesting facts about the crews.
- Draw or paint the view from a window of the ISS.

Alternative ideas? See pages 16–20

The big finish:

Create a presentation comparing and contrasting the experiences and achievements of the crews of the ISS with those of an important voyage of discovery such as Christopher Columbus (1492) or Captain James Cook (1769-70).

Alternative ideas? See pages 22–23

Fried brains?

Mobile phones – it seems everyone has one of these handy little devices these days. In fact, they are even called a 'handy' in German! There are now over a billion worldwide and numbers are growing fast. With so many people becoming reliant on these communication devices mobile phones have joined a growing group of man-made sources of electromagnetic radiation such as power lines, computers, microwaves and broadcasting transmitters that have many people worried. Harmless or dangerous? Will only time tell or does science have the answers?

Just how safe are mobile phones?

Pre-launch check:

Write down ten things you will expect to find when you do an Internet search using the words 'mobile phones dangers'. Do the search. Were you right?

Alternative ideas? See pages 13–14

Starter questions:

How do mobile phones work?

Why is there great concern about people using mobile phones?

What natural and man-made electromagnetic (EM) radiation comprise the electromagnetic spectrum?

What are the differences between high frequency ionising waves – such as x-rays – low frequency waves – such as those from electrical appliances – and radiofrequency (RF) waves – such as those from mobile phones? How do each kind of wave affect the body?

How do scientists go about working out if mobile phones are dangerous?

What is the evidence that mobile phones are safe to use?

Portfolio possibilities:

- Make a detailed diagram of the electromagnetic spectrum.
- Work with an electrician or technician to safely disassemble an unwanted mobile phone and microwave oven to identify working parts.
- Find and/or map all the mobile phone base stations in your area.
- Make an advertisement suitable for television about the dangers of using mobile phones when driving. Include statistical evidence.
- Make a mobile phone mini-museum including lots of information about how mobile phones and their features have developed.
- Find information on 20 scientific studies into the health risks of mobile phones. Group them according to the type of study, for example, laboratory studies on cells, animal experimentation and epidemiological (statistical) studies. What kind of science is involved in each?
- Compare the anxiety about mobile phones with previous worries that x-rays from colour televisions were dangerous, microwaves were a health hazard and 1980s fears that computer terminals were causing miscarriages and birth defects. Would you agree that people always worry about new technologies?

Alternative ideas? See pages 16–20

The big finish:

Interview 20 people about the use of mobile phones and identify the questions they have concerning health risks. Address these risks in a question-and-answer style pamphlet citing the best and most up-to-date scientific research.

Alternative ideas? See pages 22–23

'An air that bursts forth like the wind ... '

... was how Paracelsus described it in the sixteenth century. It makes up most of the known universe, it makes up most of you! It's colourless, odourless, tasteless and burns with an almost invisible flame. It fuels the stars and the space shuttle and one day may power your car. It has been around since the dawn of the universe. It is incredibly important stuff. It's hydrogen.

Why is hydrogen-power so important to the future of life on Earth?

Pre-launch check:

Try and fill a whiteboard with what you already know about hydrogen.

Alternative ideas? See pages 13–14

Starter questions:

What are the properties of hydrogen?

Where can hydrogen be found?

What is hydrogen currently used for?

How many different ways can we obtain hydrogen? Which ways are environmentally friendly?

How can hydrogen be used as a power source?

How do hydrogen fuel cells work and what can they be used for?

Was hydrogen responsible for the fiery death of the airship, Hindenburg?

Portfolio possibilities:

- Demonstrate a number of safe experiments that produce hydrogen. Explain what happens.
- Prepare questions and interview a scientist or engineer involved in making hydrogen-power a reality.
- Write a report on the development and use of the H-bomb.
- Ride in a hydrogen-powered vehicle, for example, Perth has some buses powered by hydrogen fuel cells. (University mechanical engineering departments may be able to help.)
- List ten problems currently encountered with using hydrogen as an energy source. For every problem find a solution.
- Write up a true-and-false quiz about hydrogen.
- Create a PowerPoint presentation about the latest hydrogen-powered cars.
- List ten ways hydrogen is being used in space exploration.
- Write an essay entitled 'We are at the end of the Oil Age and at the beginning of the Hydrogen Age'.

Alternative ideas? See pages 16–20

The big finish:

Write a letter to the Prime Minister explaining why it is vital for the government to boost research and development of hydrogen-power.

Alternative ideas? See pages 22–23

Double trouble?

What was once science fiction has become reality – breakthroughs in cloning technology are big news. Dolly, the first cloned sheep, has been followed by pigs, cows, monkeys and now, even pet cats and dogs. Scientists are successfully cloning endangered animals and even attempting to clone extinct species such as the *Thylacine* (Tasmanian tiger). The technology is available now to clone humans and one day someone will. Will the future bring headless clones harvested for body parts? People giving birth to their ancestors? Hundreds of Einsteins recreated to sort out the world's problems? Are these nightmare scenarios or hope for the sick and childless? The cloning debate is raging and we need to know what it's all about.

How does cloning work and what can it achieve?

Pre-launch check:

Make a chain of 20 identical paper dolls. Can you write a different cloning fact or sensible question on each of the dolls?

Alternative ideas? See pages 13–14

Starter questions:

How do cells work? (Make sure you understand the role of DNA, chromosomes, genes and mitosis.)

What clones occur naturally in plants and animals?

What methods have been used by scientists to produce clones?

Why is it so difficult to create a healthy clone?

What reasons do scientists have for creating animal and human clones?

What is the difference between 'therapeutic' and 'reproductive' cloning?

What ethical and moral concerns have cloning raised?

Portfolio possibilities:

- Make a model of a cell out of jelly and lollies. Use it to explain to your class how a cell works.
- Create some plant clones using tissue culture propagation.
- Construct a matrix comparing sexual reproduction, cloning and parthenogenesis.
- Create a montage of photos and headlines of Dolly the sheep and other newsworthy clones.
- Watch cell division (mitosis) under a microscope or on the Internet.
- 'Ewe was Dolly?' – Prepare and present a talk about why she was so important.
- Find out how you can now have your pet cloned (if you are rich enough).
- Find out how to keep cells alive in culture. Do it.
- Keep a log of interactive Internet sites that allow you to participate in virtual cloning and other biotechnology procedures.
- Write an essay on 'Stem cells and the possibilities of therapeutic cloning'.

Alternative ideas? See pages 16–20

The big finish:

Create a WebQuest examining the scientific and ethical issues involved in attempts to clone extinct animals.

Alternative ideas? See pages 22–23

Weapons of dark science

Advances and discoveries in science are not always used for good. They have been used to engineer deadly forms of diseases and lethal chemicals into cruel weapons. Hundreds of years ago plague-ridden corpses were catapulted into besieged castles; soldiers in World War I were blinded and killed with mustard and chlorine gas; mountains of biological weapons were stockpiled ready for slaughter during the Cold War; thousands of villagers were killed with nerve gas in northern Iraq. If these weapons were obtained by terrorists they could be coming to a neighbourhood near you.

What are the pathogens and chemicals used in Weapons of Mass Destruction (WMD) and how do they harm?

Pre-launch check:

How many biological and chemical weapons can you name?

Alternative ideas? See pages 13–14

Starter questions:

How are biological pathogens obtained and turned into weapons?

How are toxic chemicals manufactured and turned into weapons?

How, where and when have biological and chemical weapons been used?

What effects do the biological and chemical agents in these weapons have on humans and what cures or antidotes are available?

How could biological and chemical agents be delivered or dispersed to affect a large number of people?

What can be done to protect oneself from harm in the unlikely event of a chemical or bioterrorism attack?

Portfolio possibilities:

- Research the methods used to make chemical and biological weapons (bacterial culture, fermentation, electrolysis etc.). Find, perform and document experiments using these processes (with non-hazardous materials).
- Write a fictional but scientifically accurate and detailed account of either:

    a raid on a terrorist laboratory for producing biological weapons (describe the apparatus used and processes occurring in the laboratory)

    OR

    a biological or chemical terrorist attack on a city.
- Create an annotated list of informative websites (with links), articles or books about the past and present threat of chemical and biological warfare.
- Find out how gas masks work. Make your own.
- Create a 'Hall of shame' chronicling the history of chemical and biological weapons.
- Prepare a booklet for pupils at your school explaining what to do to protect themselves in the unlikely event of an attack with biological or chemical weapons.

Alternative ideas? See pages 16–20

The big finish:

Create a new game, or adapt an existing one, which uses information about the types of scientists and scientific processes used to make chemical and biological weapons and those needed to contain, treat or thwart an attack.

Alternative ideas? See pages 22–23

Mining, melting, minting marvellous metals

The Bronze Age, the Iron Age, the Computer and Space Age – the progress of human civilisation has been possible because of our growing ability to use metals to create tools, weapons, transportation and machinery. Where will metals take us tomorrow?

What are all these metals doing in my life?

Pre-launch check:

Find and attempt to answer some test questions about metals from higher year levels.

Alternative ideas? See pages 13–14

Starter questions:

What makes an element a metal?

What are the properties of metals?

How are metals 'discovered'?

Why are there more metals in use now than in the past?

How are metals mined, extracted from their ores and refined?

How are metals recycled?

Which metals are needed by our bodies? How do they contribute to our health?

How can some metals be used as germ-fighters?

What is an alloy?

Why are alloys useful?

Portfolio possibilities:

- Demonstrate every property of metals to someone who doesn't understand them.
- Enlist the help of a secondary school science teacher to experiment safely with and document:
  - reactions of 'alkali' and other metals
  - electrolysis
  - oxidisation
  - extracting copper from copper carbonate (malachite or azurite)
  - performing the alchemists 'trick' of 'turning' copper to 'silver' and then to 'gold'
- Make a series of posters about ancient metalworking technologies such as those used in jewellery, coin, tool and weapon making.
- Interview a jeweller about the appearance, qualities and uses of precious metals.
- Answer prepared questions during a visit to one or more of these places:
  - a mine or factory where metals are extracted and/or refined
  - the metallurgy department of a university
  - a metal recycling facility
- Find out how the Periodic Table is put together and how to identify metals on it.
- Find out where to find metal ores and go fossicking for some.
- Display a collection of a wide variety of metals – include information about properties and uses of each metal.

Alternative ideas? See pages 16–20

The big finish:

Become an expert about one interesting metal and use your knowledge and original ideas to generate an entry in a science competition.

Alternative Ideas? See pages 22–23

'This is the way the world ends ...'

Douglas Adams in his book *The Hitchhiker's Guide to the Galaxy* told the story of the unexpected destruction of planet Earth by a cosmic construction team clearing a path for an intergalactic freeway. Unlikely perhaps, but what does science tell us about the future of Earth? Is our planet doomed to be slowly fried by an expanding sun? Pulled apart by Nemesis, the Sun's dark twin? Swallowed by a predatory black hole? Annihilated by an asteroid? Accidentally extinguished by an out of control experiment?

What does Science tell us about the fate of planet Earth?

Pre-launch check:

List all the possible things you can imagine that could destroy Earth. (Not life on Earth but the planet itself.) When you have a really big list put your ideas in 3 categories – 'could happen now', 'could happen in the next couple of thousand years' and 'Millions or billions of years away'.

Alternative ideas? See pages 13–14

Starter questions:

How was our sun, the solar system and Earth formed?

What factors keep Earth where and how it is now?

What kinds of near-death experiences has Earth already survived?

What threats to our planet might come from space?

What is the likely future of our sun and solar system?

What, if anything, should or could we do about the destruction of the Earth?

Portfolio possibilities:

- Make a Doomsday Book – a collection of ideas and scientific speculation about how the planet might end. (Keep to scientific ideas – not religious or mystic prophecies.) Sort the ideas into categories.
- Using one of the possible scenarios write a front page story for the *Galactic Herald* about the destruction of Earth.
- Write a review of a book and a website that deals scientifically with Earth's fate.
- Make a model of an invention designed to protect Earth from one of the potential threats.
- Compare, contrast and evaluate from a scientific point of view movies that have dealt with the destruction of Earth.
- Prepare questions and interview (in person or via email) one or more scientists (preferably astrophysicists) about their views on the fate of the planet.
- Decide on the most likely scenario of how Earth will end and devise a 10-point plan (with a timetable) to ensure that the Earth's biodiversity, human civilisation and culture continue to exist elsewhere.

Alternative ideas? See pages 16–20

The big finish:

Make a documentary-style film or animation to educate your classmates about the possible futures of Earth.

Alternative ideas? See pages 22–23

The gift of life

Through scientific research and experimentation, transplants have come a long way since a Russian physician repaired a man's skull with a piece of dog's bone in the seventeenth century. Now transplants are the last best hope of thousands of critically ill people. New hearts, lungs, hands, skin, cells and maybe even faces – lifesaving body parts come from dead and living donors and sometimes from pigs and monkeys. They have been manufactured and grown in laboratories, even bought and sold on the Internet. Only one thing seems certain – we just can't get enough and that problem raises all sorts of ethical dilemmas.

How do transplants save lives and create controversies?

Pre-launch check:

Draw a picture of the human body and label all the parts that you think can be replaced. Explain how.

Alternative ideas? See pages 13–14

Starter questions:

Which organs and tissues can be successfully transplanted?

How and where are organs and other body parts obtained?

How is transplant surgery performed?

What is life like for transplant survivors?

What is xenotransplantation? Why is it controversial?

How are stem cells being used to grow new tissues and organs for transplantation?

What artificial and mechanical spare parts have been transplanted into humans?

What are the main ethical dilemmas arising out of organ and tissue transplants?

Portfolio possibilities:

- Using comic strip art tell the story of how an organ gets from the donor to the recipient.
- Should humans be using animals for spare parts? Do a S.W.O.T analysis of Xenotransplantation.
- Investigate why many people are reluctant to be listed as potential organ donors. Make an advertisement that addresses these concerns and encourages people to consider organ donation.
- Investigate the claim that poor people in some parts of the world sell their organs. Write your thoughts on whether it is ever acceptable to buy and sell body parts.
- Make a timeline about the history of heart transplants.
- Research, prepare questions and interview someone who has been the recipient of an organ donation or a medical professional involved in transplant surgery.
- Create a 'Spare Parts Catalogue' with pictures of and information about available organ and tissue parts.

Alternative ideas? See pages 16–20

The big finish:

Write a report about the latest advances and dilemmas in organ and tissue transplantation. Send it to the producers of a local children's television program or children's section of a newspaper and convince them to broadcast or publish an interesting and important aspect of your research.

Alternative ideas? See pages 22–23

Water, water everywhere and not a drop to drink

While you are reading this introduction three children will die somewhere in the world because they do not have clean, safe water to drink. Earth might be the 'Blue Planet' but only an infinitesimal percentage of the world's water is fresh and available in rivers, dams and reservoirs for use and even this tiny amount is often wasted or contaminated. Over one billion people have no clean water and five million die each year as a result. The world is facing a water crisis. The wars of the future may well be fought over access to fresh water. We desperately need to find ways to provide everyone with this precious resource.

How can science help provide clean water to everyone?

Pre-launch check:

Write a list of things that might contaminate water in developed and third world countries.

Alternative ideas? See pages 13–14

Starter questions:

Why do our bodies need water? What happens if we do not get enough?

What possible water sources are there on Earth?

What causes water shortages?

How is water contaminated and made unsafe?

What pathogens (micro-organisms that can cause disease) and contaminants are found in water and how can they make people sick?

What methods (mechanical, chemical and biological) can be used to purify and recycle water?

Portfolio possibilities:

- Find, conduct and record the results of a range of water purification experiments.
- Find out how common kitchen water filters work and what they can and can't do.
- Write an article about the pros and cons of the trend to drink bottled water.
- Prepare questions and visit your local water treatment plant. Interview a scientist who works there.
- Collect specific information about how governments and aid agencies respond to water-related problems caused by wars and natural disasters.
- Draw diagrams of recent technologies that enable water to be extracted from the air or seawater.
- Give a talk on how city-dwellers can safely harness rainwater and recycle 'grey-water'.
- Using ideas from plants or animals, come up with an innovative way of obtaining, storing or purifying water.
- Research and identify places in the world where disputes over access to fresh water are likely to happen in the near future. Consider possible solutions.

Alternative ideas? See pages 16–20

The big finish:

Liaise with an aid agency and research a real water problem affecting a third world community. Organise a display at a school function to inform the public of the problem and possible environmentally-friendly and sustainable solutions. Raise money to help make the ideas actually happen. Keep the public informed of the outcome.

Alternative ideas? See pages 22–23

More than falling apples

Ever wondered why apples fall? Why planets and stars are spheres? What causes the tides? Why we don't fall off our planet? Why the moon goes around the Earth and we all go around the Sun? Gravity is responsible for all these phenomena and although gravity is nature's 'weak' force we can never escape it – not even in space!

How does gravity affect our lives and the whole universe?

Pre-launch check:

Fill a page with drawings about how the world would be different if gravity stopped working.

Alternative ideas? See pages 13–14

Starter questions:

What is gravity?

What role does gravity play in the universe?

How did Galileo, Newton and Einstein contribute to our understanding of gravity?

What would you weigh on other planets in the solar system and why would your weight change?

Why do astronauts in space appear to be weightless?

What are the effects of the gravitational forces that exist between the Earth and the moon?

How do satellites stay in orbit?

Why is gravity considered Nature's 'weak' force?

Portfolio possibilities:

- Replicate Galileo's gravity experiments. Record what you do and what happens.
- Explain to your class how astronomers know black holes exist and what would happen to Earth if our sun became a black hole.
- Investigate how astronauts live in a microgravity situation. Draw detailed diagrams to show what adaptations you would have to make to your house if it were in orbit.
- Stand on a set of scales and go up and down in an elevator. What happens to your weight? Explain.
- Interview an astronomer about how they use their understanding of gravity to calculate the mass and position of distant objects such as stars, planets and black holes.
- Make a scrapbook of experiments that have been carried out in microgravity environments such as on board the space shuttle. Design your own, original experiment.
- Write a report about how a microgravity environment is replicated and used for training and experimentation on board special aircraft like the KC-135A.
- Make a simple accelerometer to measure 'G-forces' when you change your motion.
- Find out how to do the maths calculations necessary to launch a rocket or put a satellite in orbit.

Alternative ideas? See pages 16–20

The big finish:

Set up a mini-science centre full of hands-on activities and experiences to teach a younger group of children about gravity.

Alternative ideas? See pages 22–23

I can think a rainbow

Consider a red apple. What makes it 'red'? Would it still be red in the dark? What if you looked at it under a different coloured light – would it still be red? Would an insect looking at the apple see the same colour? If not, where did the red go? Perhaps, after all, colour is only in the mind.

What is colour?

Pre-launch check:

Can you answer the questions posed in the introduction above?

Alternative ideas? See pages 13–14

Starter questions:

What does light have to do with colour?

What causes things to appear 'coloured'?

Where do the colours that humans can see fit into the electromagnetic spectrum?

How do we actually 'see' colours? What happens in the eye and in the brain?

What is the difference between light and paint when it comes to mixing colours?

Explain the colours of the sky, the sea, plants, rainbows and bubbles.

Portfolio possibilities:

- Using prisms, investigate light and colour. Recreate Newton's famous experiment.
- Photograph a brightly coloured still life through different colour filters and using different forms of illumination. Document and explain what you discover about how the colours change.

- Design and conduct experiments that test how different coloured materials absorb light energy and convert it into thermal (heat) energy.
- Investigate what is known about the colours animals can see. Use computer software to alter photographs to appear the way different animals might see them.
- Explore optical illusions and op-art. Find out how afterimages and other visual effects are created. Using this knowledge, create your own op-art.
- Design and conduct experiments on how the colour of food and drink affects how we perceive it tastes.
- Demonstrate how infrared cameras, colour televisions or computer screens work.
- Explore, research and experiment with things that fluoresce (glow) under an ultraviolet (black) light. Document and explain your findings. (Note: 'Black lights' are considered safe but do not stare at the light or use it excessively.)
- Prepare experiments and activities for your class to learn about light and colour.
- Teach a blind child about some aspect of colour.
- Write a review of a book about the science of colour.

Alternative ideas? See pages 16–20

The big finish:

Create a room-sized interactive display about how animals and plants use and perceive colour. Consider signalling, warning, camouflage and photosynthesis.

Alternative ideas? See pages 22–23

A Shot in the dark?

In the twentieth century smallpox killed 300 million people around the world. A couple of generations ago diseases such as diphtheria, whooping cough and polio routinely killed or maimed thousands of Australian children each year. Then, vaccines to prevent these horrific diseases were hailed as medical miracles. Now, with the spectre of these killer diseases fading into the distant past, some parents wonder if vaccinations are necessary or even in the best interests of their child.

What is the evidence for and against immunisation?

Pre-launch check:

List ten ways immunisation has changed the world.

Alternative ideas? See pages 13–14

Starter questions:

How does the human body naturally protect itself against disease?

How was immunisation discovered and developed?

How do vaccinations prevent people getting sick?

What diseases have been wiped out or controlled with immunisation programs?

What is the normal vaccination schedule recommended for a child born in your country today?

What dangers have been associated with vaccinations?

Portfolio possibilities:

- Create six vaccines online in your own virtual laboratory on the NOVA website.
- Draw a comic strip illustrating how a particular vaccine is made and used and how it works.
- Prepare questions about the importance and hazards of vaccinations and interview (in person or by email) a health worker in your community and someone involved in vaccination programs in the developing world.
- Create a PowerPoint presentation on the success of immunisation programs in combating smallpox and/or polio.
- Make an annotated list of websites that provide balanced information about the role of animals in the testing and production of vaccines.
- Make a poster for parents of young children to educate them about the benefits and possible risks of immunisation. Ensure all the information on the poster is accurate, balanced and from reliable scientific sources.
- Write a report about the scientific processes involved in the development of new vaccines to prevent diseases such as cancer, AIDS, leprosy, malaria and hookworm.

Alternative ideas? See pages 16–20

The big finish:

Collect examples and make a display of inaccurate or misleading 'information' about immunisation presented by groups and individuals on the Internet. Using and quoting sources that you believe to be scientifically reputable explain why the information is wrong.

Alternative ideas? See pages 22–23

Time tourist

Imagine that instead of taking a trip to an interesting destination overseas you could stay exactly where you are but travel backwards in time thousands, millions, hundreds of millions of years? What animals would you see? Giant kangaroos, dinosaurs, trilobites, slime? How would the scenery and the climate change? Ice ages, warm ages, sea levels up and down – underwater one epoch, desert, jungle or frozen wasteland the next?

What was it like right here in the prehistoric past?

Pre-launch check:

Using a huge sheet of butcher's paper draw a timeline that goes back in time 500 million years. Add dates, drawings and information about how the planet and living things have changed.

Alternative ideas? See pages 13–14

Starter questions:

What kinds of scientists are involved in the study of the past?

What methods do scientists use to find out what happened long ago?

What is the geological time scale?

What forms of life existed during the various geological time periods?

How have the Earth's continents moved and changed in the time since life first existed?

How has the Earth's climate changed and why?

What do the major fossil discovery sites in your country reveal about life and landscapes of the past?

Portfolio possibilities:

- Visit a local museum and collect information about your region's prehistoric past.
- Prepare questions and interview a scientist whose work involves investigating the prehistoric past.
- Make a game to help you learn the order of the geological time periods and what forms of life existed in each one.
- Find ways of conveying just how long ago the geological times were – for example, if 1 year is represented by 1 mm how many metres or kilometres would you have to travel to get back to the time of the dinosaurs or to the origin of the human race?
- Visit and photograph or draw local places that reveal information about the past – for example, road cuttings with rock strata exposed, fossil sites.
- Make a dated and annotated collection of fossils (or photographs and drawings of fossils) from your region.
- Write an adventure story (using lots of actual scientific information) about a time-traveller who goes back to a particularly interesting time in your region's prehistory.
- Use a variety of art construction materials to make dioramas or murals illustrating your region's prehistory.

Alternative ideas? See pages 16–20

The big finish:

Design and create a model of a 'ride' to be installed at your local museum that takes visitors on a journey into the prehistory of your area.

Alternative ideas? See pages 22–23

Fire in the ice

It's coming from inside you, sheep and cows and from the remains of living things that are now dead. It lurks in burps, in the far corners of the solar system and under the floor of the notorious Bermuda Triangle. It can create a firestorm in a coal mine or on the surface of the sea; it can sink boats in an awful instant, heat the planet, wipe out life or provide the world with energy for hundreds of years. Meet methane.

Just what can methane do?

Pre-launch check:

Imagine methane is a person and she has agreed to speak at your school assembly. Write her speech.

Alternative ideas? See pages 13–14

Starter questions:

What are the properties of methane?

How is methane formed and where can it be found?

How does methane influence global warming?

What role do livestock, agriculture and waste disposal in landfill sites play in methane emissions?

What are methane hydrates and why are they dangerous for ships and for the planet?

In what ways can methane be hazardous to coal mines and oil rigs?

How could methane hydrates be used as a source of energy?

Portfolio possibilities:

- Examine the evidence that methane hydrates have been responsible for two mass extinction events and write up the evidence in the form of detective notes.
- Design and perform an experiment to show how bubbles of methane could sink a ship.
- Give ten reasons why the discovery of 'ice worms' living in methane hydrates is exciting. Accompany with some great pictures of the worms.
- Make a 3-minute video about scientific attempts to stop livestock burping and farting.
- Research the various ways scientists are trying to use methane as an environmentally-friendly and sustainable energy source. Prepare questions and interview scientists involved in some of these projects.
- Write a poem about the presence of methane on planets and moons in the solar system.
- Use DeBono's Six Thinking Hats to do an analysis of the idea of using methane as an energy source.
- How many disaster-movie plots can you think up that involve methane? Write some synopses.

Alternative ideas? See pages 16–20

The big finish:

Create an entertaining, informative and interactive web page about methane.

Alternative ideas? See pages 22–23.

It's not easy being green

Frogs have been around for over 200 million years. Happily hopping in soggy bogs and ponds they have seen the dinosaurs come and survived the environmental catastrophe that wiped them out. They have witnessed the super-continent Pangaea break apart and countries slip and slide over the globe. They have clung on through ice-ages and warm-ages but are now mysteriously declining, dwindling, dying, disappearing – all over the world.

What is happening to the world's frogs?

Pre-launch check:

Create a mind-map about frogs – include what you know about frog life cycles, habitats, place in the food chain and your ideas about what might be causing frogs to disappear.

Alternative ideas? See pages 13–14

Starter questions:

What is it about the bodily structure and life cycle of frogs that puts them at risk?

What habitats do frogs live in?

What role do frogs play in their habitat's web-of-life?

What is the evidence that proves frog species and populations are declining?

What theories do scientists have to explain this perplexing ecological dilemma?

Are frogs in your local area endangered?

Why is the decline and extinction of frogs important anyway?

Portfolio possibilities:

- Write a report on an endangered frog.
- Prioritise 10 things that can be done to help save frogs.
- Create an annotated, illustrated timeline of amphibian evolution.
- Classify your country's frogs on the basis of how endangered each species is.
- Prepare relevant and original questions and interview (in person or via email) a scientist who is involved in researching frog decline. Record what you have found out from them.
- Consult with local frog enthusiasts and set up a frog-breeding pond at home or school. Record and report on what happens.
- Create a MISSING poster for a frog that is now presumed extinct.
- Create a Consequences Chart – What if all the world's frogs became extinct?
- Explain this paradox – How is it that frogs survived the catastrophic upheavals that wiped out the dinosaurs but face extinction now?

Alternative ideas? See pages 16–20

The big finish:

Prepare and deliver an interactive workshop for your peers on the factors associated with the decline of frogs and the need to preserve strong and diverse frog populations.

Alternative ideas? See pages 22–23.

Not just froth and bubble

Everyone likes to play with bubbles – delightful spheres of gliding colours that grow and float and disappear into nothingness. Make enough of them and you have the strange structure we call 'foam'. But foams are not all made of soap – there are metal foams, plastic foams, rubber, ceramic, chocolate, rock and frog-egg foams just to name a few. Eat them, sleep on them, use them to put out fires, keep your drinks cold, protect you head in an accident or shave your face – foam is truly amazing stuff!

What is the science behind bubbles and foams?

Pre-launch check:

Imagine you own a shop that sells every kind of foam imaginable. Write a list of what you sell.

Alternative ideas? See pages 13–14

Starter questions:

What are the properties of bubbles?

How do bubbles form and change over time?

What foams can be found in nature, food, drink, toiletries and cosmetics, industry, building, packaging, fire fighting, furniture, medicine, sports and transport?

What are some of the many ways foams can be made?

What types of scientists study bubbles and foams and why?

What kinds of foams will be useful in future technologies?

Portfolio possibilities:

- Choose five interesting foams that you encounter in your everyday life. Explore their properties, find out about their manufacture, alternative uses and disposal.
- Create a mobile about foams in flight – from the structure of bird bones to the insulation on the space shuttle.
- Write an illustrated report on the use of bubbles and foams by animals such as frogs, fish and insects to protect themselves, their eggs and their young.
- Make a PowerPoint presentation about what happens to water films, bubbles and foams in microgravity.
- Find and try experiments with bubbles and foams. Record and report the results.
- Investigate different formulae and methods to produce record-breaking soap bubbles.
- Design and conduct your own experiments to test the sound or heat insulation properties of a variety of foams.
- Find out how to make antibubbles and demonstrate the procedure to your class.

Alternative ideas? See pages 16–20

The big finish:

Become a food scientist. Find out about foams in food – honeycomb, mousse, meringue, sponge cakes, bread, marshmallows etc. Design and conduct experiments to find the best methods and ingredients to create the lightest foams. Create new recipes using foams and write a foam foods cookbook.

Alternative ideas? See pages 22–23.

Earth's security blanket

Imagine what it would be like if Earth had no atmosphere – no air to breathe, no blue sky, no weather and no warmth. We would be blasted by cosmic radiation and rained on by meteorites. Our planet would be waterless and dead.

How does the atmosphere support life on Earth?

Pre-launch check:

Write one thing you know about the atmosphere for every letter of the word – A-T-M-O-S-P-H-E-R-E.

Alternative ideas? See pages 13–14

Starter questions:

Why does Earth have an atmosphere? How did it form?

What is the atmosphere made of?

How are the layers in the atmosphere different?

How does energy from the Sun interact with the atmosphere?

What causes weather?

What are the 'natural' and 'enhanced' greenhouse effects?

How has human activity altered the atmosphere?

Why do things 'burn up' as they come through the atmosphere?

How would things be different if Earth had no atmosphere?

Portfolio possibilities:

- Find out what ozone is and what it does in the atmosphere. Find and conduct experiments to detect the presence and effects of 'bad ozone.'
- Make a scale model of the Earth's atmosphere, placing planes, clouds, balloons etc. in their correct position. (Do the International Space Station and satellites orbit within the atmosphere?)
- Research and present a tribute detailing the bravery and achievements of high-altitude balloon pioneers entitled, 'Going to Great Heights'.
- Using charts and graphs compare Earth's atmosphere to those of the other planets in the Solar System.
- Make a multimedia presentation about what happens to spacecraft, satellites and meteors as they plunge down through the atmosphere.
- Compose a soundtrack based on the man-made and natural events that occur in different layers of the atmosphere.
- Write a report on how mountain climbers or sportspeople are affected by high altitudes.
- Explain the phenomena of auroras and make a bulletin-board display of relevant information and photographs.

Alternative ideas? See pages 16–20

The big finish:

Write and illustrate a book about an alien species that fouled their atmosphere with industrial pollutants and upset their planet's delicate balance with disastrous results.

Alternative ideas? See pages 22–23

It's a worm's world after all

'Worms' are what we call lots of different kinds of animals that look like … well … worms. There are earthworms – Charles Darwin thought they were possibly the most important animals on the planet. They can grow up to 7 m long and were the only living things to survive the crash of the space shuttle, Columbia. There are worms that live in ponds and seas, in ice and even in hot water. Unfortunately, there are also lots of gross and dangerous ones that live inside people and animals. (Finding out about these ones will really make you squ-worm!)

What are the world's worms doing?

Pre-launch check:

Tapeworms in humans can grow up to 10 m long. Starting with that fact, use a long strip of paper and see how many metres of worm facts you can write.

Alternative ideas? See pages 13–14

Starter questions:

The word 'worm' refers to many different taxonomic groups of animals. Choose one worm from each of these groups and describe its lifecycle, anatomy, nervous system, reproductive system and method of locomotion.

- flatworms (Platyhelminthes)
- ribbon worms (Nemertea)
- roundworms (Nematoda)
- segmented worms (Annelida)

How do earthworms benefit the soil and agriculture?

What worms infest humans?

How are humans infested by worms?

What are the consequences of worm infestations in humans and what can be done about the problem?

Portfolio possibilities:

- Find out how to extract nematodes from soil samples and observe them with a magnifying glass or microscope. Draw what you see.
- Make an illustrated glossary of worm terms and become an expert in their meanings.
- Find pictures of different worms and group them according to their taxonomy.
- Prepare questions and interview a vet about worm infestations in animals and a doctor about worm infestations in humans.
- Visit a museum and interview a palaeontologist about worm fossils.
- Make a poster of tough worms that survive in extreme habitats.
- Write an illustrated report about the use of leeches in medicine.
- Learn how to recognise the following worms in pond water samples – oligochaetes, flatworms and nematodes. Use a microscope to find them.
- Make a 'monster gallery' of 2D and 3D art based on microscopic pictures of worms.
- Enlist the assistance of a scientist who specialises in marine worms and organise an excursion to investigate seashore worms.

Alternative ideas? See pages 16–20

The big finish:

Become an amateur oligochaetologist. Research, observe, perform and document experiments with earthworms. Set up and continuously improve a worm farm that effectively composts your classes' vegetable scraps and experiment with growing plants in the resulting compost. Publish your findings in the school newsletter.

Alternative ideas? See pages 22–23

To eternity and beyond

Death is part of life and always has been but what actually happens when someone dies? What happens to the body and why? Can decomposition of the body be postponed or even prevented – dried out like an Egyptian mummy, pickled in a cask of brandy like Admiral Horatio Nelson or frozen like Otzi the Iceman? Some people in their quest for immortality put their hope in Science and their bodies in the deep freeze awaiting 'resuscitation' while others send their ashes into space. Death and dealing with it – fascinating stuff.

What does science know about death?

Pre-launch check:

Can you come up with 100 words associated with death?

Alternative ideas? See pages 13–14

Starter questions:

How do we know death has occurred?

What natural processes happen to the body of a human or animal after death?

How have some bodies been accidentally preserved? (for example, bog bodies, Incan 'mummies')

How have some bodies been deliberately preserved? (for example, Egyptian mummies, Lenin)

What is cryogenics? Could it enable dead bodies to be resuscitated?

How is the preservation of bodies similar to methods used to preserve food?

What options are there for the disposal of bodies? Can they be recycled?

Portfolio possibilities:

- Visit a museum and find dead humans and animals there. Collect information on how the bodies have been preserved and how different cultures deal with death.
- Create a display with examples of 'mummies' from all over the world.
- Prepare questions and interview one or more people who deal with dead bodies, for example, undertaker, forensic entomologist, a coroner or doctor.
- Prepare a pamphlet outlining and explaining modern options for disposing of bodies.
- Collect controversies regarding death and prepare and conduct surveys to assess public opinion.
- Design a round-the-world tour to places where death is dealt with in interesting ways.
- Debate the issue, 'Science should be searching for a way to make us immortal'.
- Observe the decomposition of a dead animal. Record what happens.
- Investigate the science behind the work of fiction, 'Frankenstein'. Could life ever be created from body parts?
- Investigate the claim that people executed by guillotine remained conscious for a time after losing their heads or that some animals like chickens and cockroaches can 'live' without a head.

Alternative ideas? See pages 16–20

The big finish:

You are an alien scout from Planet Valhalla where everyone is now immortal. Explain to your fellow Valhallans (actually your classmates) how the mortal humans deal with death.

Alternative ideas? See pages 22–23

May the force be with you

Thousands of years ago people discovered rare stones with 'magical' powers. These stones could mysteriously attract pieces of iron and when allowed to swing on the end of a string always pointed north. Once only useful for finding one's way, today magnets are used in hundreds of different ways – everything from sticking your artwork on the fridge to levitating trains, from scanning your brain to cleaning oily seabirds. The Earth itself is one big magnet and there are even tiny magnets inside your brain. Magnets, magnetism, magnetic fields – can you feel the force?

What is this force called magnetism?

Pre-launch check:

Collect 10 different questions from friends or family about magnetism. How many can you answer?

Alternative ideas? See pages 13–14

Starter questions:

Are magnets made, found or created? How?

How does a magnet work?

How does a compass work?

What are magnets used for?

What is a magnetic field?

How is the Earth like a huge magnet?

What does the Earth's magnetic field do?

How do animals make use of Earth's magnetic field?

Portfolio possibilities:

- Find and/or design and conduct experiments involving magnetism. Record and share your findings.
- Magnets are everywhere but we are usually not aware of them – find out where all the magnets in our everyday life are lurking.
- Write a report on how a particular animal uses its 'magnetic sense'.
- Find out why scientists are predicting an imminent and potentially disastrous reversal of the Earth's magnetic field. Prepare questions and interview at least one scientist working on this problem.
- Collect examples of toys that use magnetism.
- Use magnets to collect micrometeorites.
- Explain to an audience how the Earth's magnetic field creates the spectacular auroras. Illustrate your talk with lots of good pictures.
- Compare and contrast the properties of a regular magnet and an electromagnet.
- Design a game to teach others about magnetism.

Alternative ideas? See pages 16–20

The big finish:

Design and build an invention that uses magnets for an original and practical purpose. Enter your invention in a science competition.

Alternative ideas? See pages 22–23

'Milk's leap toward immortality'

- Clifton Fadiman, 1957

Perhaps accidentally discovered when a nomadic herdsman's milk stored in a 'bottle' made from a sheep's stomach turned solid after a morning trek on a hot day, cheese has been a nutritious and delicious part of our diet for many thousands of years. Today there are over 4000 varieties worldwide including yak, reindeer, buffalo, camel and zebra cheese – all made with the help of friendly micro-organisms and a little 'cheese science'. You can even make your own!

What is the science behind cheese-making?

Pre-launch check:

Imagine you are abducted by some cheese-loving aliens and will only be released if you tell them how cheese is made. What do you tell them?

Alternative ideas? See pages 13–14

Starter questions:

How is cheese made? Explain what happens during the major stages in cheese-making.

Why is temperature important in cheese-making?

What do the following cheese-making words mean: lactose, rennet, protein, casein, pasteurisation, homogenisation, enzyme, curds, whey.

What is the role of micro-organisms (bacteria and moulds) in cheese-making?

Why is hygiene so important in cheese-making?

What are the main types of cheese and how is each type made?

What happens when you 'age' cheese? How is it done?

Portfolio possibilities:

- Visit a cheese-making factory and/or a cottage industry cheese-maker to see how cheese is made.
- Yoghurt making is an easy introduction to some of the techniques used in cheese-making. Experiment with making some. (Be very careful with hygiene. Enlist the help of a responsible adult.)
- Find out how to make cheese at home. Try cottage, mozzarella, ricotta, mascarpone, neufchatel, labneh and paneer for starters. (Be very careful with hygiene. Enlist the help of a responsible adult.) Record what you try in a cheese-making diary.
- Organise a cheese-tasting and information session. Explain how the different cheeses are made.
- Observe bacterial cultures from yoghurt under a microscope.
- Imagine there was a 'zoo' where 'friendly' and dangerous micro-organisms found in food were kept. Design a visitors' guide to the creatures in this zoo.
- Prepare questions and interview a scientist involved in cheese production.
- Design and conduct an experiment to determine if the flavour of cheese improves if served at room temperature.

Alternative ideas? See pages 16–20

The big finish:

Demonstrate cheese-making to your class, explaining the science of the processes as you work.

Alternative ideas? See pages 22–23

A burning question

Humans are the only species to use fire. Since we first discovered how to use it over half a million years ago we have worshipped it, cooked with it, warmed ourselves with it, made tools with it, chased wild animals away with it; used it to shape our environment, to propel us, to provide light in the dark and to toast marshmallows. But – do we really know what it is?

How has our understanding of and ability to use fire changed over time?

Pre-launch check:

Explain fire (scientifically) to a creature from another planet where fire does not exist.

Alternative ideas? See pages 13–14

Starter questions:

What is fire?

What ideas about fire did people have in the past?

What causes a fire to start?

What actually happens when something burns?

What are the different types of fires?

Is there a difference between burning and oxidation?

What makes some things more flammable than others?

How do fireproof materials and fire extinguishers work?

Portfolio possibilities:

- Find out how people have made fire from prehistoric times until the invention of the match, and working safely with a responsible adult, experiment with and document attempts to make fire as these people did.
- Combine three-dimensional art and an understanding of chemistry and explain to your class how fireworks work.
- What kinds of scientists work on investigating, using and preventing fires? List and explain their jobs.
- Interview a scientist or firefighter about the factors that contribute to the intensity and destructiveness of a bushfire.
- Research the importance of French scientist, Antoine Lavoisier's (1743–94) work in discovering the true nature of fire. Role-play the character of Lavoisier on the eve of his execution and in a monologue explain his contribution to science.
- Collect and sort information about spontaneous combustion into categories and use the information to make hazard posters for three different workplaces.
- Research spontaneous human combustion. What scientific explanations are there for this strange phenomenon?
- Give a fire safety demonstration to your class.
- Make a model of a blast furnace and explain how it works.

Alternative ideas? See pages 16–20

The big finish:

Create an entertaining 30-minute quiz and variety show for your class about fire.

Alternative ideas? See pages 22–23

Are you what you eat?

Most of us eat when we are hungry or because we are used to meals at a certain time but what about people who eat as a sport? How about 432 oysters in ten minutes? What happens to the food that slips out of sight down your throat and what goes wrong when you vomit? Do you know someone who gets really sick when they eat a particular food? Why do some people like to eat live monkey brains or poisonous fish or even swallow swords? Why do some people eat dirt and some cows eat rabbits and what do you do when your dog swallows a plastic bag? You will need a strong stomach for this challenge.

How do we use and abuse food?

Pre-launch check:

From memory, draw a labelled diagram of the digestive system explaining what each part does.

Alternative ideas? See pages 13–14

Starter questions:

How does the digestive system work in humans?

How do the digestive systems of birds, fish, insects and reptiles differ from that of humans?

What causes vomiting and diarrhoea?

What amount and balance of food is considered a 'healthy diet' for someone your age?

What are the causes and symptoms of malnutrition?

What is the connection between the brain and eating? Investigate the following medical conditions: anorexia, bulimia and pica.

What happens when someone is allergic to a particular food?

Portfolio possibilities:

- Find and try some experiments that relate to the digestion of food. Record what you do and what you find out.
- Report on how care workers help malnourished children back to health.
- Present an audiovisual presentation on the pros and cons of the new 'sport' – competitive eating.
- Make a poster about the prevention and treatment of stomach 'bugs' or eating disorders.
- Research the claims made about 'live bacteria' yoghurt drinks and report on your findings.
- Interview a veterinarian about eating disorders and digestive problems in pets.
- Keep an annotated scrapbook of news items relating to food science.
- Keep a food journal for a week. Work out ten ways to improve your diet.
- Write a report about five science careers associated with food and/or digestion.
- How does 'sword swallowing' work? Give an illustrated talk on this phenomenon.
- Imagine you are travelling to an isolated base in the Antarctic for a year and will need to take all your food with you. Work out your dietary requirements and draw up a list of suitable food supplies for a whole year.

Alternative ideas? See pages 16–20

The big finish:

Write and produce a play for a younger class that introduces them to the amazing world of eating.

Alternative ideas? See pages 22–23

Pull the other one

Science is all about exploring and trying to understand the unknown and mysterious. Because of this, there has always been and probably always will be people who believe, state or 'discover' something in the realm of science that is not true. Claims are made that involve fake fossils, prehistoric giants, man-bats on the moon, perpetual motion, alien autopsies and human clones. The media pick up the stories and many people can be fooled until scientific methods finally expose the lie.

What do scientific hoaxes tell us about the ideas of the time and how scientists find the truth?

Pre-launch check:

Log-on to a reputable science forum and start a discussion using all the facts you already know about scientific hoaxes and frauds. Save a copy for your teacher.

Alternative ideas? See pages 13–14

Starter questions:

What are ten well-known examples of scientific hoaxes or deceptions?

What fields of science have been particularly prone to hoaxes?

What kinds of people have perpetrated scientific hoaxes? What motivated them?

How were the hoaxes discovered or revealed?

What methods do scientists use to find out the truth about things?

Portfolio possibilities:

- Research three interesting hoaxes from three different centuries. Explain how the hoaxes fit into the context of the scientific thought of the time. For example, Piltdown Man was unearthed at a time when scientists were very keen to find the 'missing link' between humans and our ape-like ancestors.
- Write a 500-word essay on the topic, 'Scientific hoaxes succeed because people want to believe them'.
- Red Faces – make five portraits of scientists who have been famously fooled. Add information to each portrait explaining how they were deceived.
- Using solid research and convincing acting and effects, make a video of a science news report, suitable for April Fools Day, about a scientific breakthrough or disaster.
- Hall of Shame – make a corridor display of 'wanted' posters revealing what a variety of pranksters, hoaxers and untrustworthy scientists have got up to.
- Keep a scrapbook of examples of how mass-media perpetrates unfounded and incorrect scientific ideas e.g. sensationalist reporting of miracle cures or imminent environmental catastrophes. How can you find out if a report is true or not?
- Write a story about a scientific hoax that goes horribly wrong.

Alternative ideas? See pages 16–20

The big finish:

Some people claim that the lunar landings of the 1960s and 70s were fake. Examine the arguments and counterarguments and present the evidence and your conclusions in the form of a courtroom trial.

Alternative ideas? See pages 22–23

Over the moon

Cheese, cows, werewolves, lunatics and legends – the moon has been worshiped and studied, sung to and speculated about for thousands of years. When Galileo observed its mountains with a newly invented telescope in 1610 science gradually began to dispel the myths. In 1959 a spaceship from Earth crashed into it and in 1969 six hundred million people watched as humans walked on it. The moon tugs on our oceans. It slows the Earth. Things just wouldn't be the same without it.

How would life on Earth be different if there was no Moon?

Pre-launch check:

Using the Internet or substantial books about the moon, write out a list of words associated with the Moon. For example, waxing, crescent, lunar module, lunar eclipse, mare, crater. Using only your current knowledge try and write definitions for each of the words.

Alternative ideas? See pages 13–14

Starter questions:

How was the moon formed?

How and why does the moon's appearance change?

Why do lunar eclipses occur?

What kinds of rocks can be found on the moon? Where or how did they originate?

How have humans acquired information about the moon?

What is the evidence for water existing on the moon? Why does it matter?

How do the moon and Earth influence each other?

Portfolio possibilities:

- Compare and contrast our moon with another interesting moon in the solar system.
- Create a timeline that shows how scientific understanding of the moon has developed.
- The moon has no atmosphere – list 10 fascinating consequences.
- Create an annotated display of photographs and quotes from the Apollo program.
- Research Australia's role in the exploration of the moon. Prepare questions and interview a scientist who has been involved.
- Some people claim that the Apollo moon landings were a hoax. Research their arguments and determine if they really have a case.
- Organise a 'Moon Day' for your class when everyone comes to school dressed as one of the people who have been honoured by having a lunar feature named after them.
- Make a glossary of interesting expressions related to the moon, for example, 'Once in a blue moon,' 'Harvest Moon' etc. Explain the science behind the terms.
- Create a moving model that demonstrates why the moon's appearance changes.
- Examine old science fiction books or films for references to people landing on the moon. What did writers and/or filmmakers get right and wrong?

Alternative ideas? See pages 16–20

The big finish:

After researching the benefits gained from past lunar missions and considering the future needs of humankind, take part in a debate on the topic, 'It is vital that the government should invest in a swift return to the moon'.

Alternative ideas? See pages 22–23

When clouds collide

There are few things as fascinating, awesome and scary as lightning. Lightning is spectacular and terrifying; hotter than the sun and ear-burstingly loud. It kills hundreds of people every year and can strike 'out of the blue'. Ancient people imagined that lightning bolts were the weapons of the gods and even today lightning is not fully understood and myths and misinformation abound.

How does lightning work and what can it do?

Pre-launch check:

Write a sentence to explain each of these terms:
lightning, thunder, electrons, protons, electricity, static electricity, charge, voltage, ionized air, leaders, streamers, return stroke, sprite, lightning rod, fulgurite.

Alternative ideas? See pages 13–14

Starter questions:

What causes lightning?

What is lightning?

What is thunder?

What is 'static electricity' and how is it related to lightning?

What different kinds of lightning exist? (Include rare and newly discovered types.)

How are people injured by lightning?

How do you keep safe in a thunderstorm?

Portfolio possibilities:

- Find and/or design experiments to explore static electricity. Record your results.
- Borrow, buy or make a VandeGraaff generator and use it to conduct experiments.
- Research survivors' stories to find out what it is like to be struck by lightning.
- Create a visually interesting notice board display of your classmates' questions about lightning and your well researched answers.
- Draw a series of diagrams to explain what lightning would look if you could slow it down enough for the human eye to see what is going on.
- Write a 'Weird Science' or 'Freaky Phenomena' article about St Elmo's Fire and Ball Lightning for your school newsletter.
- Design and make 'Lightning Safety Information' packs for a variety of outdoor sporting organisations in your area.
- Write a report about Benjamin Franklin's famous kite and key experiment.
- Check out a museum for a specimen of a fulgurite (petrified lightning).
- Use lightning maps for your country to determine which areas experience high, moderate and low lightning activity.

Alternative ideas? See pages 16–20

The big finish:

Create an entertaining but scientifically accurate comic book about lightning. Include the science behind lightning as well as lightning myths explained, stories, things to do, safety tips and … Dr Frankenstein.

Alternative ideas? See pages 22–23

'Smell is a potent wizard'

- Helen Keller

Flies do it with their feet; moths do it with their antennae; spiders do it with their hairy legs and we do it with our noses and our brains. It is the way, as newborns, that we instinctively recognise our mothers; it can attract us unconsciously to others; transport us back in time to things or places we had forgotten about; give us intense pleasure; affect our mood or signal danger. Smell — our forgotten and often misunderstood sense.

How does our sense of smell work and why is it important?

Pre-launch check:

Draw a labelled diagram that depicts what you know about how your sense of smell works.

Alternative ideas? See pages 13–14

Starter questions:

How does the sense of smell work in humans and in a variety of different animals?

In what ways do humans and animals rely on their sense of smell?

How can a smell make us vividly remember something from the past?

How is our sense of taste related to our sense of smell?

How are perfumes, artificial smells like 'scratch and sniffs' and room fresheners made?

What are the latest ideas and inventions that use odours and the sense of smell?

Can smells make you sick?

Portfolio possibilities:

- Draw a detailed and labelled diagram of the olfactory system.
- Prepare questions and interview a neurologist about disorders of the olfactory system and diseases linked to our sense of smell.
- Conduct research and experiments to find out how well people can identify smells and what links there are between our senses of taste and smell.
- Prepare and conduct a survey to research people's favourite smells – does age or gender affect our choices?
- Give a five-minute talk on the latest ideas to use smells in electronic games, at the cinema and over the Internet.
- Design, conduct and report on experiments that determine the effectiveness of aromatherapy.
- Compare and contrast the human sense of smell with that of five different animals.
- Write an article about the work of cadaver dogs for a children's science magazine.
- Create an annotated list of informative websites about the sense of smell.

Alternative ideas? See pages 16–20

The big finish:

Write an essay entitled, 'Smell – our forgotten and misunderstood sense' and have it read and commented upon by a neurologist or other smell expert.

Alternative ideas? See pages 22–23

Crystal clear

Named after the Greek word for 'ice', lightly dusting your fish and chips or sweetening your tea, set in a king's crown, in the snout of a trout, in a gouty toe, in your ear, in computers and every form of electronic equipment, falling as snow – intricate, useful, beautiful – common and rare – crystals are everywhere.

What are all these crystals doing in my life?

Pre-launch check:

Create a mind-map of what you know about crystals.

Alternative ideas? See pages 13–14

Starter questions:

What is a crystal?

What can crystals be made of?

How do crystals grow and form in nature?

How are crystals grown artificially?

What types (categories) of crystals are there?

What do crystallographers do?

Where can crystals be found?

What can crystals be used for?

Portfolio possibilities:

- Explore gem collections in museums, books, shops and on the Internet. Go prospecting or collect real gemstones and/or photographs and learn to identify different types.
- Explain and demonstrate how piezoelectricity works and make a display of devices that make use of this special property of some crystals.
- Find out about X-ray crystallography. Visit a laboratory to see how it is done and interview a scientist about the importance and applications of this technology.
- Prepare questions and interview a jeweller about the range and value of gemstones. Try to arrange to see gemstones being cut.
- Find out how and why crystals are being grown on the International Space Station.
- Make a piece of art based on a patchwork of microscopic images of crystals.
- Snowflakes are crystals that form from water vapour. Find a way to make some.
- Make a crossword about the connection between crystals and volcanoes.
- On a life-size drawing of a human show where crystals can be found inside the body.
- Explain why pseudo-science claims that crystals can 'heal' are incorrect.
- Build your own crystal radio.
- Find out how to observe crystals growing under a microscope.

Alternative ideas? See pages 16–20

The big finish:

Grow your own crystals. Try variations in method and materials to see what effects are produced. Be scientific about it – questioning, hypothesising, controlling variables, experimenting, recording etc. Present examples of your crystals and a summary of what you have found out as an entry in a science competition.

Alternative ideas? See pages 22–23

The chicken and the egg

You probably know lots about chickens, tadpoles and caterpillars and how they grow and change into adult animals and in time produce their own young but what about planulas, phyllasomas or nymphs? What do they become? How long do they live? A few weeks like some fruit flies? A hundred years like some turtles or thousands of years like some species of trees? And what makes living things grow old and die anyway? Life cycles take many unique and fascinating forms.

What are some of nature's most fascinating life cycles?

Pre-launch check:

Make a chart detailing the life cycles of these living things: humans, dragonflies, lady beetles, mushrooms, barnacles, oak trees, jellyfish, koalas, sharks and crabs. Include stages of development and approximate time periods for each stage. Check references to see how well you did.

Alternative ideas? See pages 13–14

Starter questions:

At what point does life begin? Is it the same for all living things?

What makes living things grow and stop growing?

What makes living things age and die?

What are some of the many different ways living things reproduce?

Why do some living things have very long life cycles while others are very brief?

Portfolio possibilities:

- Select and grow or breed a variety of suitable animals and plants that have brief life cycles. Observe, measure and document their development.
- Research and create diagrams of ten really unusual life cycles.
- Create and perform a rap or a dance about the life cycle of a living thing that interests you.
- Write and perform a puppet play about the life cycles of parasites that infect humans.
- Make up a matching game using 'before' and 'after' pictures of insects that undergo metamorphosis.
- Find out about the life cycles of animals and plants that are pests and determine what makes them so 'successful'. Compare with the life cycles of animals or plants that are facing extinction.
- With others, brainstorm lots of interesting questions about life cycles and find the answers. For example, what does a baby jellyfish look like? Which animal lives the longest? What animal is pregnant the longest? How old is a cockroach when it reproduces? Do all insects lay eggs?

Alternative ideas? See pages 16–20

The big finish:

Using suitable images and music but no words, make a meaningful audiovisual presentation about life cycles.

Alternative ideas? See pages 22–23

Dead in the water?

Picture these scenes:

- the Snowy Mountains Scheme where 25 years of engineering marvels and hard work turned a river around and sent it flowing into Australia's dry interior
- irrigation channels surging like arteries of life-giving water across the dusty land
- windmills lifting water from the Great Artesian Basin to make the desert bloom
- mighty dams that provide the people and industries of cities with dependable water.

Are they symbols of progress, human ingenuity and achievement or evidence of short-sighted and arrogant disregard for the balance of nature?

How can we protect and sustain our precious water resources?

Pre-launch check:

Draw a detailed 'before and after' picture of how Australia's waterways have been exploited and changed since European settlement.

Alternative ideas? See pages 13–14

Starter questions:

How does the water cycle work?

How much rainfall does different parts of Australia receive and where does it all go?

Why are rivers, wetlands and estuaries important?

What problems threaten our river systems and who or what is responsible?

What is 'groundwater'? How is it used? Why is it under threat?

What causes 'dry land salinity' and why is it such a problem?

What is being done to protect and manage water resources in a sustainable way?

Portfolio possibilities:

- Assess and test the health of a local waterway with a Waterwatch group.
- Visit and photograph a waterway that is under threat from pollution or algae blooms. Identify the cause of the problem.
- Draw a series of pictures showing the effects of development and urbanisation over time on a city's original waterway such as Sydney's Tank Stream.
- Make a documentary about the use and misuse of the Great Artesian Basin.
- List five ways higher sea levels due to global warming will affect water supplies.
- Write a poem about past and present effects of mining or industry on waterways.
- Make an audiovisual PowerPoint presentation about the role of scientists in monitoring, restoring and sustaining the health of waterways.
- 'We dump less waste in rivers now than in the past.' Support or refute this statement with evidence.
- Make an entertaining and educational board game about the problems affecting river systems and groundwater.
- Create a play about the pros and cons of irrigation or dams.

Alternative ideas? See pages 16–20

The big finish:

You have been zapped back to 1949 and have been appointed chief advisor to the government that is about to embark on the Snowy Mountains Scheme. Consider the issues and needs of the country in 1949 and what we know now about the positive and negative consequences of the Scheme. In a dramatic presentation, explain how and why the plan should be changed.

Alternative ideas? See pages 22–23

Walking, creeping, crawling and flying with dinosaurs

Insects – creeping, crawling, slithering, scurrying, climbing, sucking, biting, infesting, flying, dancing and building for the last 400 million years or so. Where did they come from? Were they all giants like the metre-wide dragonfly? Did dinosaurs have fleas? And what came first – flowers or bees? Welcome to the fascinating world of prehistoric bugs.

What is the insects' story?

Pre-launch check:

Complete an A–Z of insect facts by writing the alphabet down one side of a large sheet of paper and writing one thing you know about insects for each letter.

Alternative ideas? See pages 13–14

Starter questions:

What makes an insect an insect – is it just about having six legs?

What are the different kinds (orders) of insects?

How do insects breathe, eat and reproduce?

What senses do insects have and how do they use them?

How do scientists know about prehistoric insects?

How have insects changed in the last 400 million years?

Why are insects so successful?

Portfolio possibilities:

- Compare and contrast five prehistoric insects with their modern counterparts.
- Take photographs of insects. Identify and categorise them according to the order they belong to, e.g. Diptera (flies), Coleoptera (beetles) etc.
- Write a report about and make a model of an interesting prehistoric insect.
- Explore insect fossils at a museum. Draw sketches, take notes and interview a scientist who works with the collection.
- Find out about the fascinating work of entomologists and paleoentomologists. Prepare questions and interview one of these scientists about their work.
- Make fake 'fossils' of insects using dead insects and plaster of paris.
- Find out how insects can be preserved in amber and what information modern science can extract from these preserved animals.
- Make a collection of weird but true facts about insects.
- Write a story called 'A Bug's Life' but set it during the Carboniferous period.
- Create a detailed and informative mural that tells the story of the evolution of insects.

Alternative ideas? See pages 16–20

The big finish:

Using displays of ancient and modern insects that can be found in your local museum, create a series of interesting worksheets for museum visitors to complete. Present them to the curator of the museum.

Alternative ideas? See pages 22–23

Any port in a storm?

Walls of water capsizing and sinking ships; towering waves breaking through shops and houses and sweeping over the land; hundreds of thousands of people drowning; ferocious winds snapping poles, uprooting trees and blowing whole towns away. Welcome to the most feared weather phenomenon of all – the tropical cyclone.

What are tropical cyclones and why are they so dangerous?

(Note: Tropical cyclones are also known as 'hurricanes' and 'typhoons'.)

Pre-launch check:

How long you can talk about tropical cyclones without deviating from the topic or repeating yourself.

Alternative ideas? See pages 13–14

Starter questions:

What are tropical cyclones?

Where, when and how are tropical cyclones formed?

What factors can influence the size, intensity, path and death of a tropical cyclone?

How do cyclones create 'storm surges' and what damage can these surges do?

How is information about tropical cyclones gathered and relayed around the world?

Are global warming and tropical cyclone activity linked?

Portfolio possibilities:

- Research and find out what scale is used to measure the intensity of tropical cyclones and find photographs of and news reports about each kind.
- Make a labelled model of a tropical cyclone and use it in a presentation about how cyclones work.
- Prepare questions and interview a meteorologist about how they forecast and track cyclones.
- Compose a pamphlet detailing precautions that should be taken during cyclone season and what to do to keep safe if a cyclone actually threatens.
- Compile a list of tropical cyclone records, for example, strongest winds, largest, most loss of life, longest lasting.
- Present an illustrated lecture on how the forecasting of cyclones has changed from pre-wireless days to recent developments in satellite technology.
- Keep records of, and track and report on a tropical cyclone as it happens.
- Prepare questions and interview someone who has lived through a destructive cyclone.
- Find out what attempts have been made to destroy tropical cyclones and, using your knowledge of tropical cyclones, come up with a different idea.

Alternative ideas? See pages 16–20

The big finish:

Research the history (frequency, magnitude and damage) of a cyclone-prone region that interests you. Design a cyclone-proof tourist resort that will withstand the destructive forces of a severe cyclone. Justify your design.

Alternative ideas? See pages 22–23

They may be little but ...

They were the first and only kind of life on Earth for 2 billion years. They put the oxygen in the atmosphere that allowed all other life to evolve and they are still with us – just about everywhere from oceans, ponds, flower pots and drains to deserts, rocks and polar ice. They make flamingos pink and polar bears green. They enable plants to grow and provide our food ... but ... they can turn waterways into choked cauldrons of stinking toxic slime and even kill you. Meet the Cyanobacteria Family. Don't underestimate them.

Why are cyanobacteria so important?

(Note: Cyanobacteria are sometimes known as blue-green algae but this is misleading because they are not algae at all.)

Pre-launch check:

C-Y-A-N-O-B-A-C-T-E-R-I-A. Can you write one fact about cyanobacteria for each letter?

Alternative ideas? See pages 13–14

Starter questions:

What exactly are cyanobacteria?

Where do cyanobacteria fit into the Kingdoms of Life?

How do cyanobacteria live and reproduce?

Where do cyanobacteria occur?

What role do cyanobacteria play in their environments and in the global ecosystem?

Cyanobacteria are often called 'blue-green algae'. How are cyanobacteria similar and different to other kinds of microscopic life such as real (eukaryotic) algae?

Why do dangerous 'algal blooms' occur and what can be done to prevent them?

Portfolio possibilities:

Note – some cynaobacteria are toxic! Do not touch, inhale or eat them!

- Prepare questions and interview a scientist who is working with cyanobacteria.
- Draw a detailed diagram that illustrates how cyanobacteria created the oxygen in our atmosphere.
- Make a slideshow of images of cyanobacteria.
- Some cyanobacteria form structures called stromatolites. Visit living stromatolites or find fossilised ones in a museum or at a university.
- Visit the site of an 'algal bloom' and make a photographic record. Find out the cause of the bloom, how to fix the problem and how to prevent it from occurring again.
- Make a model of a waterway affected by an 'algal bloom' and use it to teach your class about the factors that contribute to the problem.
- Conduct experiments to test the influence of fertilisers on algal blooms.
- Make up and act out a T.V. game similar to 'Perfect Match' based on the symbiotic relationships that cyanobacteria have with other organisms.
- Make a radio show about cyanobacteria that exist in extreme habitats and the implications for finding life elsewhere in our solar system.

Alternative ideas? See pages 16–20

The big finish:

Make several Winogradsky Columns using organic matter from different environments. Observe and record the microbial growth for at least a month. Study the organisms under a microscope and find out how to identify them. Explain the functioning of these complex mini-ecosystems to your class.

Alternative ideas? See pages 22–23

Hubble, Hubble, toil and trouble

Since being launched into Earth orbit in 1990 the Hubble Space Telescope has provided us with spectacular images of galaxies and planets. It has shown us the birth and death of stars and proved the existence of black holes. It has looked 12 billion years into the past and completely changed our understanding of the universe. It was never meant to last forever but in January 2004, NASA announced (to the shock and dismay of scientists and space enthusiasts worldwide) that the Hubble Space Telescope will be abandoned.

Is the Hubble Space Telescope worth saving?

Pre-launch check:

How good is your knowledge of astronomy? Find some online astronomy quizzes and try them. Make a note of your strengths and weaknesses.

Alternative ideas? See pages 13–14

Starter questions:

What is the Hubble Space Telescope and how does it work?

What can the Hubble Space Telescope do?

How has the Hubble Space Telescope been kept functioning in space?

In what ways are the Space Shuttle Program and the Hubble Space Telescope linked?

What are the major achievements of the Hubble Space Telescope? Why are they important?

How has the Hubble Space Telescope changed our view of our place in the universe?

Portfolio possibilities:

- Find out if you can see the Hubble Space Telescope pass over from where you live. If so, observe its transit.
- Make a detailed model of the Hubble Space Telescope. Use the model to explain to your class how Hubble works.
- Research the importance of gyroscopes to the Hubble Space Telescope. Make your own gyroscope.
- Interview a scientist who has used information supplied by the Hubble Space Telescope about what more could be achieved if the telescope was given extra time.
- Write a poem comparing the Hubble Space Telescope to a time-traveller.
- Explain why the achievements of the Hubble Space Telescope are sometimes compared to those of the Italian astronomer, Galileo.
- Give your class weekly updates on what the Hubble Space Telescope has been exploring.
- Make an annotated collection of images captured by Hubble.
- Write an obituary for the Hubble Space Telescope.

Alternative ideas? See pages 16–20

The big finish:

Create a web page about the importance of the Hubble Space Telescope. Use some original graphics and include links to great Hubble websites.

Alternative ideas? See pages 22–23

Air now – gone tomorrow?

It's rare, it's blue, it smells like a thunderstorm, it can make you sick or even kill you but life on Earth cannot survive without it. Some scientists are desperate to get rid of it; others are desperate to get it back. The whole world is talking about it – ozone.

What is ozone and how can it be both good and bad?

Pre-launch check:

Try an ozone quiz on the Internet.

Alternative ideas? See pages 13–14

Starter questions:

What is ozone?

How is ozone formed in the stratosphere?

How does the ozone in the stratosphere protect life on Earth?

What has gone wrong with the 'ozone layer'?

What consequences will the depletion of the 'ozone layer' have on life on Earth?

How is 'bad' or ground-level ozone formed?

What harm does 'bad' ozone do?

Portfolio possibilities:

- Use online air-quality data available from government environmental protection agencies to monitor and record ozone levels that contribute to pollution in your local area or nearest city. Determine what factors contribute to fluctuations in ozone levels.
- Plan and carry out an investigation using an Ecobadge (available online from Vistanomics) to determine ozone levels in your school, home and local environment.
- Millions of workers are exposed to damaging levels of ozone in the workplace. Investigate this issue and do something about it.
- Use satellite data available online from NASA's Total Ozone Mapping Spectrometer to track and record changes in the density of the ozone layer over the Antarctic.
- Write and perform a puppet play about a superhero determined to save Earth's primary radiation shield – the ozone layer.
- Find and photograph sources of ozone-depleting chemicals in your home and neighbourhood. Create an annotated slide show with your photographs.
- Find out if there is a link between the thinning of the ozone layer and an increase in skin cancers. Write a list of recommendations for individuals and for the government.

Alternative ideas? See pages 16–20

The big finish:

Find a web site or brochure advertising an ozone generator (air purifier). Examine the claims the advertisement makes about ozone and determine which of these are scientifically correct and which are misleading or wrong. Share what you have found out with a consumer information organisation.

Alternative ideas? See pages 22–23

The long and winding road

Ever wondered why there are so many different kinds of living things? Why do whales seem a lot like sharks but feed their babies milk just like humans? Where were the people when the dinosaurs were around? What happened to the mammoths? Why do chimpanzees seem so much like us? What will humans look like a million years from now? Interesting questions and the answers are all about a process called 'evolution'.

How has life on Earth evolved?

Pre-launch check:

Draw a picture of what the ancestors of humans looked like five million, 50 million, 500 million and a billion years ago.

Alternative ideas? See pages 13–14

Starter questions:

What is evolution?

How does 'natural selection' lead to the development of new species?

What do genes do and what does genetic mutation have to do with evolution?

How are all living things related to each other?

Who was Charles Darwin and what led him to develop his ideas about evolution?

Why is evolution a controversial theory?

What evidence supports the theory of evolution?

Portfolio possibilities:

- Evolution has many opponents. Collect examples of 'evidence' used to disprove evolution and discuss this 'evidence' with a scientist. Write your conclusions.
- Read and review a book about Charles Darwin.
- Invent a game to teach your classmates about natural selection.
- Learn how traits and appearances are passed on to new generations via genes and demonstrate your understanding using Punnett Squares.
- Create illustrated time lines of the evolution of a variety of animals.
- Create a 'tree' diagram showing the evolution of humans.
- Write a story about Darwin time-travelling to the present and discovering how modern science has continued to find evidence to support his ideas.
- Look at your own family tree to see how appearances and traits have been passed down from one generation to the next. Make a poster of your findings.
- Visit a fossil collection in a museum and/or interview a palaeontologist or geologist to find out how scientists work out the age of fossils and the Earth itself.
- Write a response to the question, 'If humans evolved from apes, why are there still apes in the world?'

Alternative ideas? See pages 16–20

The big finish:

Antibiotic-resistant 'superbugs', pesticide-resistant pests, new influenza viruses; swiftly-evolving microbes are outsmarting humans and causing real problems. Decide on 20 practical ideas to help humans beat the bugs before it is too late. Communicate your ideas to someone able to do something about the problem.

Alternative ideas? See pages 22–23.

Resources

Consequences chart

A consequence chart shows the sequence of events or stages of a plan. For example:

Starting event	Subsequent event	subsequent event	Final event

Edward de Bono's six thinking hats

A resource to aid thinkers to conduct a richer, more balanced exploration of any subject. An individual or group can put on one of the coloured hats to perform a specific thinking operation, as the hat serves as a tool to guide the thinking process.

White Hat: Information - what are the facts?

- Red Hat: Feelings - what do I feel about this?
- Black Hat: Judgment - what is wrong with this?
- Yellow Hat: Benefits - what are the good points?
- Blue Hat: Thinking - what thinking is needed?
- Green Hat: Creativity - what new ideas are possible?

For further information and resources regarding this topic please refer to Edward de Bono (1992), *Six thinking hats for schools (books 1-4)*. Victoria: Hawker Brownlow Education.

Mind map

A clear visual tool to represent a mental model of the topic under focus. For example:

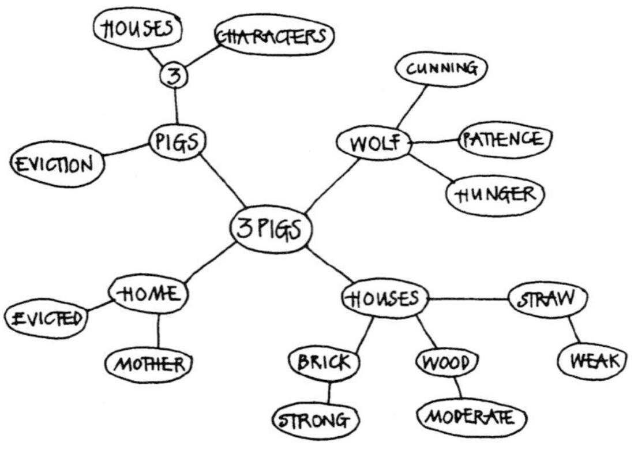

For further information on using mind maps please refer to Oliver Caviglioli, Ian Harris and Bill Tindall (2004), *Thinking skills and eye Q*. Victoria: Hawker Brownlow Education.

PMI analysis

A way of delaying conclusions by forcing you to look at optional viewpoints. Developed by Edward de Bono.

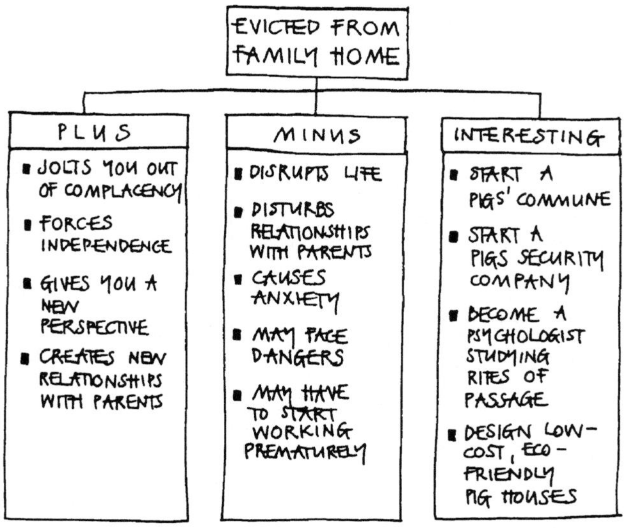

For further information on using PMI please refer to Oliver Caviglioli, Ian Harris and Bill Tindall (2004), *Thinking skills and eye Q*. Victoria: Hawker Brownlow Education.

SCAMPER

Substitute, Combine, Adapt, Modify (Magnify, Minify), Put to another use, Eliminate and Reverse. Developed by B. Eberle (1991) as an acronym for a range of problem solving techniques that may be applied in any creative situation.

S	Substitute	Substitute a person or object, or serve in a different way or role.
C	Combine	Bring together or unite Blend ideas, materials, situations.
A	Adapt	Adjust to suit another purpose or set of conditions.
M	Modify Magnify Minify	Modify attributes, frequency or size. Enlarge to make larger in form or quality. Reduce to make smaller, lighter, less frequent.
P	Put to another use	Use for an alternative purpose, in another situation or in a different way.
E	Eliminate	Remove or omit a quality, a part or the whole.
R	Reverse	Change order, adjust or create a new sequence, layout or scheme.

For further information on using SCAMPER please see the following:

B. Eberle (1991), *Scamper - Games for imagination development.* Victoria: Hawker Brownlow Education.

Michael Pohl (2000), *Learning to think - thinking to learn.* Victoria: Hawker Brownlow Education.

SWOT analysis

Strenghts, Weaknesses, Opportunities and Threats. A way of graphically representing the various positive and negative aspects of projects or organisations.

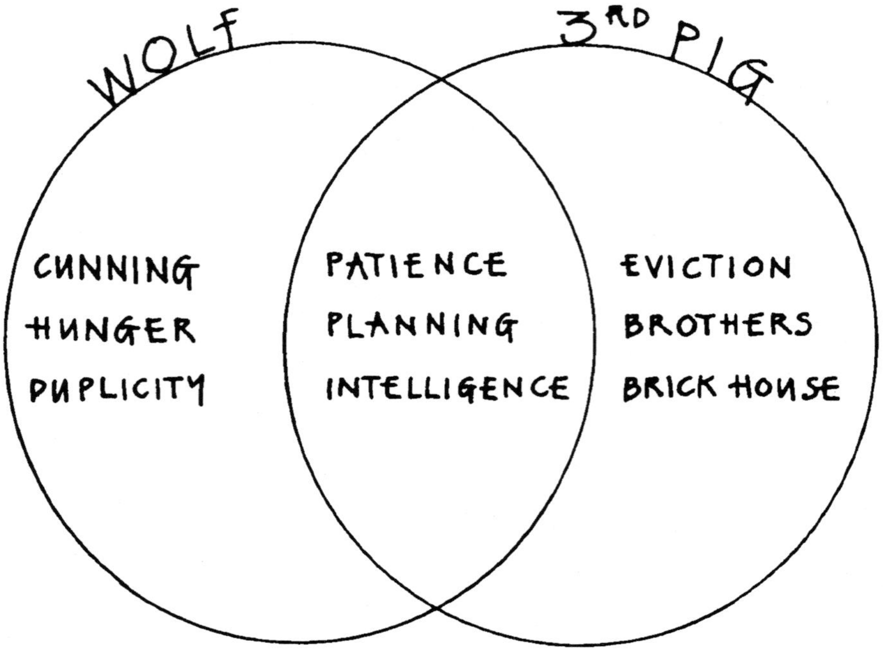

For further information on using S.W.O.T please refer to Oliver Cavigliolo, Ian Harris and Bill Tindall (2004), *Thinking skills and eye Q. Victoria*: Hawker Brownlow Education.

Edward de Bono's Six Thinking Hats

Winogradsky columns

A strategy developed by Sergei Winogradsky to isolate organisms from nature in a miniature model pond cross section. Instructions for constructing Winogradsky columns can be found in many laboratory manuals.